LEADERSHIP and the NEW SCIENCE

LEADERSHIP and the NEW SCIENCE

Learning about Organization from an Orderly Universe

MARGARET J. WHEATLEY

Berrett-Koehler Publishers
San Francisco

Berrett-Koehler Publishers, Inc.
155 Montgomery St.
San Francisco, CA 94104-4109

Ordering Information

Orders by individuals and organizations. Berrett-Koehler publications are available through bookstores or can be ordered direct from the publisher at the Berrett-Koehler address above or by calling (800) 929-2929.

Quantity sales. Berrett-Koehler publications are available at special quantity discounts when purchased in bulk by corporations, associations, and others. For details, write to the "Special Sales Department" at the Berrett-Koehler address above or call (415) 288-0260.

Orders by U.S. trade bookstores and wholesalers. Please contact Prima Distribution, P.O. Box 1260, Rocklin CA 95677-1260; tel. (916) 786-0426; fax (916) 786-0488.

Orders for college textbook/course adoption use. Please contact Berrett-Koehler Publishers. 155 Montgomery St., San Francisco, CA 94104-4109; tel. (415) 288-0260; fax (415) 362-2512.

Printed in the United States of America

Printed on acid-free and recycled paper that meets the strictest state and U. S. guidelines for recycled paper (50 percent recycled waste, including 10 percent postconsumer waste).

Library of Congress Cataloging-in-Publication Data

Wheatley, Margaret J.
Leadership and the new science: learning about organization from an orderly universe / Margaret J. Wheatley. — 1st ed.
p. cm.
Includes bibliographical references and index.
ISBN 1-881052-01-X (hardcover: alk. paper)
1. Science. 2. Organization. 3. Quantum theory. 4. Self-organizing systems. 5. Chaotic behavior in systems. I. Title.
Q158.5.W43. 1992
500—dc20. 92-70095
CIP

First Edition
First Printing 1992

To my family

Contents

ILLUSTRATIONS

"The world is richer than it is possible to express in any single language."

—Ilya Prigogine

Preface

I like to think of this book as reminiscent of the early chart books used by explorers sailing in search of new lands. Those early maps and accompanying commentary were descriptive but not predictive, enticing but not fully revelatory. They pointed in certain directions, illuminated landmarks, warned of dangers, yet they included enough elusive references and blank spots to encourage explorations and discoveries by other people. They contained colorful embellishments of areas or events that had struck the wanderer's fancy and ignored other, important places. They contained life-saving knowledge, passed hand to hand among those who were willing to dare similar voyages of their own.

This book contains the charts of my first journeys into the territory of new science—those hypotheses and discoveries in biology, chemistry, and physics that challenge us to reshape our fundamental world view. I wandered in this new land as a stranger, one who had been trained not in science, but in organizational theory and practice. The more I explored, the louder grew the siren's song. This new territory contained powerful images, metaphors, and ways of thinking that asked me to seek new ways of comprehending the issues that trouble organizations most: chaos, order, control, autonomy, structure, information, participation, planning, and prediction.

My guides were the many books written for lay readers over the past twenty years that describe an array of scientific discoveries and formulations that date back to the turn of the century. That was the first stage of the journey—extensive reading and the clear but nascent notion that I was in a land rich in possibilities.

The second stage was the process of writing this book. I watched as ideas and shapes took form, created by this new relationship between me and my organizational experiences and the ideas of science. The quantum world teaches that there are no pre-fixed, definitely describable destinations. There are, instead, potentials that will form into real ideas, depending on who the discoverer is and what she is interested in discovering. Only by venturing into the unknown do we enable new ideas to take shape, and those shapes are different for each voyager.

I have drawn my charts for any and all who work in or are affected by life in organizations, and who feel, as I do, that there must be a better way to organize work, people, and life. In the hope that you, too, might become excited by these new ways of seeing, I report on and translate the new science discoveries that I found significant. As a reporter, I have been biased by my own training. I have focused on the scientific discoveries that intrigued my organizational mind and have ignored many others. This is neither a comprehensive nor a technical guide to new science. It recounts, instead, the trips that I took to but a few of the emerging areas in science, those that beckoned most strongly to me.

In addition to recounting the science that intrigued me, I have pointed out interesting places where science could potentially inform organizational theory and action, areas that deserve more serious exploration, perhaps by you. But there were also areas that called out loudly for my attention, and for these I have begun to spin out implications and connections that intrigue me.

So there are maps of many varieties in this book. Some describe certain new science findings, hopefully in enough detail that you get a sense of their terrain.

Others point out areas for further inquiry, without knowing yet what will be found there. Still others are more detailed, drawing deliberate connections between science and organizations. And, finally, there are personal records of my own journey, recounting my feelings and experiences as I brought back questions and treasures and applied them to my own management practice.

Three different branches of science are treated in some detail: quantum physics, self-organizing systems, and chaos theory. Because I develop the science as I go along, things will make more sense if you read the chapters in order. The Introduction and chapter 1 introduce all three sciences and the new contributions they make to our understanding of the way the world works. They also provide some initial explanations of sources of order in the universe and speculations on the fears and conditioning that prevent us from appreciating the way that order is created in complex systems.

Chapters 2 through 4 explore the implications of quantum physics for organizational designs and practices that have, until now, been predicated on the seventeenth-century world view of Sir Isaac Newton. Quantum physics challenges our thinking about observation and perception, participation and relationships, and the influences and connections that are created across large and complex systems.

The next chapters, 5 and 6, deal with the scientific theory of self-organizing or dissipative structures. These chapters introduce new ways of understanding disequilibrium and change, as well as the uses of disorder in creating new possibilities for evolutionary growth. Information, in this view of the self-organizing universe, is the primal energy that structures matter into form, the necessary ingredient for continued life. Self-organizing structures also demonstrate new relationships between autonomy and control, showing how a large system is able to maintain its overall form and identity only because it tolerates great degrees of individual freedom.

Chaos theory is the subject of chapter 7. Given a world where chaos and order exist in tandem, where stability is never guaranteed but chaos always conforms to a boundary, I propose my own hypothesis for the forces in organizations that create the structured shape that holds up through chaotic times. I also explore lessons to be learned from fractals—how nature creates its diverse and intricate shapes by enumerating a few basic principles and then permitting great amounts of autonomy.

This is a book about the early stirrings of new ways of thinking about organizations. As such, it does not lend itself to definitive conclusions. But in chapter 8 and the Epilogue, I attempt to draw together various principles from the new science and propose a "new" scientific approach to management. There are several critical management issues that would be better served by explanations from the new science, rather than the old science to which we have clung. But to explore these discoveries in more depth will require many years, new relationships among us all, and a new openness to inquiry. In the Epilogue, I comment on my own experience with the process of discovery.

Having made these early forays, I know that I am changing in deep and evolving ways. I hope that you will venture even farther and return with gifts of your own.

Mapleton, Utah
March 1992

Margaret J. Wheatley

The Author

Margaret J. Wheatley is associate professor of management at Brigham Young University, a consultant to organizations large and small, and co-founder of The Berkana Institute. This book is both a radical departure from her past work, and a natural synthesis of the diverse wanderings of her life.

Her early plans for a scientific career ended at age nineteen, when she switched to studies of history and English literature at the University of Rochester and University College London. Two years in the Peace Corps in Korea led to several years as an educational administrator of programs in low-income communities in New York. She became a student of systems theory and communications during graduate work at New York University, where she received her M. A. in communications. It was there that she discovered the field of organizational behavior, which led her to Harvard University. She received her doctorate from Harvard in the Program for Administration, Planning, and Social Policy, with a primary focus on organizational diagnosis and interventions.

Wheatley's consulting career began as a founding member of Rosabeth Kanter's firm, Goodmeasure, Inc. Several years later, she co-founded Ibis Consulting Group, Inc. in Cambridge, Massachusetts. She has consulted to a wide variety of Fortune 500 clients, educational, and non-profit institutions, working at all levels—from CEOs to assembly line workers. Most recently, she has been involved in participative techniques that involve the whole system of an organization in planning desired organizational futures. She also speaks frequently to corporate groups and professional associations on creating workplaces of opportunity and meaning, and on designing organizational structures and cultures that can sustain rapid change.

In 1990, she co-founded The Berkana Institute, a non-profit organization, to help facilitate communities of inquiry and support among those who are experimenting both with new organizational forms and with managing from a more integrated sense of self.

Her publications include a book on work and family issues, as well as several articles on the impact of opportunity and power in large organizations, the creation of ethical work, and the motivating power of ethics in times of corporate confusion.

She and her husband, Nello-John Pesci, together have seven children and are recent East-coast emigrés to Utah, where, in addition to their full-time work, they run a small horse ranch and fruit orchard.

Acknowledgments

One day, as I was seated at my desk writing, surrounded by stacks of books, I experienced a deep sense of gratitude for all the writers who had taken time to search and think and relate their insights. I couldn't have learned what I did without their efforts and commitment, and I admire their discipline and curiosity. My first acknowledgment of thanks is to the authors listed in the bibliography.

For my own writing venture, I want to express my deep appreciation to the following people for the help they provided in the form of insights, encouragement, and good thinking: Lavina Fielding Anderson, Maurice Atkin, Rachel Bassett, Jim Bell, Paul Carlile, Sandra Claudell, J. Bart Czirr, Gil Dube, Sarah Eames, Badi Foster, John Grassi, Samuel Guider, Jill Kanter, Reba Keele, Myron Kellner-Rogers, Gary L. Jensen, Grant Lasson, Henry Lester, Eileen Morgan, Randy Moore, Steven Piersanti, Larry Rees, Jaye Sacks, Lisa Vincent, Marvin Weisbord, Lana Wertz, and Dale Wright. And especially to my big, bustling, loving, and always supportive family.

"To my mind there must be,
at the bottom of it all, not an equation,
but an utterly simple idea.
And to me that idea, when we finally discover it,
will be so compelling, so inevitable,
that we will say to one another,
'Oh, how beautiful.
How could it have been otherwise?'"

—John Archibald Wheeler

INTRODUCTION

Searching for A Simpler Way to Lead Organizations

I am not alone in wondering why organizations aren't working well. Many of us are troubled by questions that haunt our work. Why do so many organizations feel dead? Why do projects take so long, develop ever-greater complexity, yet so often fail to achieve any truly significant results? Why does progress, when it appears, so often come from unexpected places, or as a result of surprises or serendipitous events that our planning had not considered? Why does change itself, that event we're all supposed to be "managing," keep drowning us, relentlessly reducing any sense of mastery we might possess? And why have our expectations for success diminished to the point that often the best we hope for is staying power and patience to endure the disruptive forces that appear unpredictably in the organizations where we work?

These questions had been growing within me for several years, gnawing away at my work and diminishing my sense of competency. The busier I became with work and the more projects I took on, the greater my questions grew. Until I began a journey.

Like most important journeys, mine began in a mundane place—a Boeing 757, flying soundlessly above America. High in the air as a weekly commuter between Boston and Salt Lake City, with long stretches of reading time broken only by occasional offers of soda and peanuts, I opened my first book on the new science—Fritjof Capra's *The Turning Point*, which described the new world view emerging from quantum physics. This provided my first glimpse of a new way of perceiving the world, one that comprehended its processes of change and patterns of connections.

I don't think it accidental that I was introduced to a new way of seeing at 37,000 feet. The altitude only reinforced the message that what was needed was a larger perspective, one that took in more of the whole of things. From that first book, I took off, seeking out as many new science books as I could find in biology, evolution, chaos theory, and quantum physics. Discoveries and theories of new science called me away from the details of my own field of management inquiry and up to a vision of the inherent orderliness of the universe, of creative processes and dynamic, continuous change that still maintained order. This was a world where order and change, autonomy and control were not the great opposites that we had thought them to be. It was a world where change and constant creation signalled new ways of maintaining order and structure.

I don't believe I could have grasped these ideas if I had stayed on the ground.

During the past fifteen to twenty years, books that translate new science findings for lay readers have proliferated, some more reputable and scientific than others. Of the many I read, some were too challenging, some were too bizarre, but others contained images and information that were breathtaking. I became aware that I was wandering in a realm that created new visions of freedom and possibility, giving me new ways to think about my work. I couldn't always draw immediate corollaries between science and my dilemmas, but I noticed myself developing a new serenity in response to the questions that surrounded me. I was reading of chaos that contained order; of information as the primal, creative force; of systems that, by design, fell apart so they could renew themselves; and of invisible forces that structured space and held complex things together. These were compelling, evocative ideas, and they gave me hope, even if they did not reveal immediate solutions.

Somewhere—I knew then and believe even more firmly now—there is a simpler way to lead organizations, one that requires less effort and produces less stress than the current practices. For me, this new knowledge is only beginning to crystallize into applications, but I no longer believe that organizations are inherently unmanageable in our world of constant flux and unpredictability. Rather, I believe our present ways of understanding organizations are skewed, and that the longer we remain entrenched in our ways, the farther we move from those wonderful breakthroughs in understanding that the world of science calls "elegant." The layers of complexity, the sense of things being beyond our control and out of control, are but signals of our failure to understand a deeper reality of organizational life, and of life in general.

We are all searching for this simplicity. In many different disciplines, we live today with questions for which our expertise provides no answers. At the turn of the century, physicists faced the same unnerving confusion. There is a frequently told story about Niels Bohr and Werner Heisenberg, two founders of quantum theory. This version is from *The Turning Point*:

> In the twentieth century, physicists faced, for the first time, a serious challenge to their ability to understand the universe. Every time they asked nature a question in an atomic experiment, nature answered with a paradox, and the more they tried to clarify the situation, the sharper the paradoxes became. In their struggle to grasp this new reality, scientists became painfully aware that their basic concepts, their

> language, and their whole way of thinking were inadequate to describe atomic phenomena. Their problem was not only intellectual but involved an intense emotional and existential experience, as vividly described by Werner Heisenberg: "I remember discussions with Bohr which went through many hours till very late at night and ended almost in despair; and when at the end of the discussion I went alone for a walk in the neighboring park I repeated to myself again and again the question: Can nature possibly be so absurd as it seemed to us in these atomic experiments?"
>
> It took these physicists a long time to accept the fact that the paradoxes they encountered are an essential aspect of atomic physics. . . . Once this was perceived, the physicists began to learn to ask the right questions and to avoid contradictions . . . and finally they found the precise and consistent mathematical formulation of [quantum] theory.
>
> . . . Even after the mathematical formulation of quantum theory was completed, its conceptual framework was by no means easy to accept. Its effect on the physicists' view of reality was truly shattering. The new physics necessitated profound changes in concepts of space, time, matter, object, and cause and effect; and because these concepts are so fundamental to our way of experiencing the world, their transformation came as a great shock. To quote Heisenberg again: "The violent reaction to the recent development of modern physics can only be understood when one realizes that here the foundations of physics have started moving; and that this motion has caused the feeling that the ground would be cut from science." (In Capra 1983, 76-77. Used with permission)

For the past several years, I have found myself often relating this story to groups of managers involved in organizational change. The story speaks with a chilling authority. Each of us recognizes the feelings this tale describes, of being mired in the habit of solutions that once worked yet that are now totally inappropriate, of having rug after rug pulled from beneath us, whether by a corporate merger, reorganizations, downsizing, or a level of personal disorientation. But the story also gives great hope as a parable teaching us to embrace our despair as a step on the road to wisdom, encouraging us to sit in the unfamiliar seat of not knowing and open ourselves to radically new ideas. If we bear the confusion, then one day, the story promises, we will begin to see a whole new landscape, one of bright illumination that will dispel the oppressive shadows of our current ignorance. I still tell Heisenberg's story. It never fails to speak to me from this deep place of reassurance.

I believe that we have only just begun the process of discovering and inventing the new organizational forms that will inhabit the twenty-first century. To be responsible inventors and discoverers, though, we need the courage to let go of the old world, to relinquish most of what we have cherished, to abandon our interpretations about what does and doesn't work. As Einstein is often quoted as saying: No problem can be solved from the same consciousness that created it. We must learn to see the world anew.

There are many places to search for new answers in a time of paradigm shifts. For me, it was appropriate that my inquiry led back to the natural sciences, reconnecting me to an earlier vision of myself. At fourteen, I aspired to be a space biologist and carried thick astronomy texts on the New York subway to weekly classes at the Hayden Planetarium. These texts were far too dense for me to understand, but I carried them anyway because they looked so impressive. My abilities in biology were better founded, and I began college with full intent to major in biology, but my initial encounters with advanced chemistry ended that career, and I turned to the greater ambiguity of the social sciences. Like many social scientists, I am at heart a lapsed scientist, still hoping that the world will yield up its secrets to me in predictable formulations.

But my focus on science is more than a personal interest. Each of us lives and works in organizations designed from Newtonian images of the universe. We manage by separating things into parts, we believe that influence occurs as a direct result of force exerted from one person to another, we engage in complex planning for a world that we keep expecting to be predictable, and we search continually for better methods of objectively perceiving the world. These assumptions, as I explain in chapter 2, come to us from seventeenth-century physics, from Newtonian mechanics. They are the base from which we design and manage organizations, and from which we do research in all of the social sciences. Intentionally or not, we work from a world view that has been derived from the natural sciences.

But the science has changed. If we are to continue to draw from the sciences to create and manage organizations, to design research, and to formulate hypotheses about organizational design, planning, economics, human nature, and change processes (the list can be much longer), then we need to at least ground our work in the science of our times. We need to stop seeking after the universe of the seventeenth century and begin to explore what has become known to us in the twentieth century. We need to expand our search for the principles of organization to include what is presently known about the universe.

The search for the lessons of new science is still in progress, really in its infancy, but what I hope to convey in these pages is the pleasure of sensing those first glimmers of a new way of thinking about the world and its organizations. The light may be dim, but its potency grows as the door cracks wider and wider. Here there are scientists who write about natural phenomena with a poetry and a lucidity that speak to dilemmas we find in organizations. Here there are new images and metaphors for thinking about our own organizational experiences. This is a world of wonder and not knowing, where scientists are as awestruck by what they see as were the early explorers who marvelled at new continents. In this realm, there is a new kind of freedom, where it is more rewarding to explore than to reach conclusions, more satisfying to wonder than to know, and more exciting to search than to stay put.

This is not a book of conclusions, cases, or exemplary practices of excellent companies. It is deliberately *not* that kind of book, for two reasons. First, I no longer believe that organizations can be changed by imposing a model developed elsewhere. So little transfers to, or even inspires, those trying to work at change in their own organizations. Second, and much more important, the new physics cogently explains that there is no objective reality out there waiting to reveal its secrets. There are no recipes or formulae, no checklists or advice that describe "reality." There is only what we create through our engagement with others and with events. Nothing really transfers; everything is always new and different and unique to each of us.

This book attempts to be true to that new vision of reality, where ideas and information are but half of what is required to evoke reality. The creative possibilities of the ideas represented here depend on your engagement with them. I have interpreted my task as presenting material to provoke and engage you, knowing that your experience with these pages will produce different ideas, different hopes, and different experiments than did mine. It is not important that we agree on one expert interpretation or one sure-fire application. That is not the nature of the universe in which we live. We inhabit a world that is always subjective and shaped by our interactions with it. Our world is impossible to pin down, constantly changing and infinitely more interesting than we ever imagined.

Though the outcomes to be gained from reading this book are unique to each reader, the ideas I have chosen to think about focus on the meta-issues that concern those of us who work in large organizations: What are the sources of order? How do we create organizational coherence, where activities correspond to purpose? How do we create structures that move with change, that are flexible and adaptive, even boundaryless, that enable rather than constrain? How do we simplify things without losing both control and differentiation? How do we resolve personal needs for freedom and autonomy with organizational needs for prediction and control?

The new science research referred to comes from the disciplines of physics, biology, and chemistry, and from theories of evolution and chaos that span several disciplines. Each chapter inquires into metaphorical links between certain scientific perspectives and organizational phenomena, but it may be useful first to say something in general about the directions of new science research.

Scientists in many different disciplines are questioning whether we can adequately explain how the world works by using the machine imagery created in the seventeenth century, most notably by Sir Isaac Newton. In the machine model, one must understand parts. Things can be taken apart, dissected literally or representationally (as we have done with business functions and academic disciplines), and then put back together without any significant loss. The assumption is that by comprehending the workings of each piece, the whole can be understood. The Newtonian model of the world is characterized by materialism and reductionism—a focus on things rather than relationships and a search, in physics, for the basic building blocks of matter.

In new science, the underlying currents are a movement toward holism, toward understanding the system as a system and giving primary value to the relationships that exist among seemingly discrete parts. Donella Meadows, a systems thinker, quotes an ancient Sufi teaching that captures this shift in focus: "You think because you understand *one* you must understand *two*, because one and one makes two. But you must also understand *and*" (1982, 23). When we view systems from this perspective, we enter an entirely new landscape of connections, of phenomena that cannot be reduced to simple cause and effect, and of the constant flux of dynamic processes.

Explorations into the subatomic world began early in this century, creating the dissonance described in Heisenberg's story. In physics, therefore, the search for radically new models now has a long and somewhat strange tradition. The strangeness lies in the pattern of discovery that characterized many of the major discoveries in quantum mechanics. "A lucky guess based on shaky arguments and absurd ad hoc assumptions gives a formula that turns out to be right, though at first no one can see why on earth it should be" (March 1978, 3). I delight in that statement of scientific process. It gives me hope for an approach to discovery that can influence the methodical, incremental, linear work that leads to the plodding character of most social science research.

The quantum mechanical view of reality strikes against most of our notions of reality. Even to scientists, it is admittedly bizarre. But it is a world where *relationship* is the key determiner of what is observed and of how particles manifest themselves. Particles come into being and are observed only in relationship to something else. They do not exist as independent "things." Quantum physics paints a strange yet enticing view of a world that, as Heisenberg characterized it, "appears as a complicated tissue of events, in which connections of different kinds alternate or overlap or combine and thereby determine the texture of the whole" (1958, 107). These unseen *connections* between what were previously thought to be separate entities are the fundamental elements of all creation.

In other disciplines, especially biology, the use of nonmechanistic models is much more recent. At the outer edges of accepted practice (although gaining slowly in credibility) are theories like the Gaia hypothesis, which sees the earth as a living organism actively engaged in creating the conditions which support life, or Rupert Sheldrake's morphogenic fields, which describe species memory as contained in invisible structures that help shape behavior. Some of what we know how to do, Sheldrake argues, comes not from our own acquired learning, but from knowledge that has been accumulated in the human species field, to which we have access. Whole populations of a species can shift their behavior because the content of their field has changed, not because they individually have taken the time to learn the new behavior.

So many fundamental reformulations of prevailing theories in evolution, animal behavior, ecology, and neurobiology are underway that, in 1982, Ernst Mayr, a noted chronicler of biological thought, stated: "It is now clear that a new philosophy of biology is needed" (1982, 73).

In chemistry, Ilya Prigogine won the Noble Prize in 1977 for his work demonstrating the capacity of certain chemical systems (dissipative structures) to regenerate to higher levels of self-organization in response to environmental demands. In the older, mechanistic models of natural phenomena, fluctuations and disturbances had always been viewed as signs of trouble. Disruptions would only more quickly bring on the decay that was the inevitable future of all systems. But the dissipative structures that Prigogine studied demonstrated the capacity of living systems to respond to disorder (non-equilibrium) with renewed life. Disorder can play a critical role in giving birth to new, higher forms of order. As we leave behind our machine models and look more deeply into the dynamics of living systems, we begin to glimpse an entirely new way of understanding fluctuations, disorder, and change.

New understandings of change and disorder are also emerging from chaos theory. Work in this field, which keeps expanding to take in more areas of inquiry, has led to a new appreciation of the relationship between order and chaos. These two forces are now understood as mirror images, one containing the other, a continual process where a system can leap into chaos and unpredictability, yet within that state be held within parameters that are well-ordered and predictable.

New science is also making us more aware that our yearning for simplicity is one we share with natural systems. In many systems, scientists now understand that order and conformity and shape are created not by complex controls, but by the presence of a few guiding formulae or principles. The survival and growth of systems that range in size from large ecosystems down to tiny leaves are made possible by the combination of key patterns or principles that express the system's overall identity and great levels of autonomy for individual system members.

The world described by new science is changing our beliefs and perceptions in many areas, not just in the natural sciences. I see new science ideas beginning to percolate in my own field of management theory. One way to see their effect is to look at the problems that plague us most in organizations these days or, more accurately, what we *define* as the problems. Leadership, an amorphous phenomenon that has intrigued us since people began studying organizations, is being examined now for its relational aspects. More and more studies focus on followership, empowerment, and leader accessibility. And ethical and moral questions are no longer fuzzy religious concepts but key elements in our relationships with staff, suppliers, and stakeholders. If the physics of our universe is revealing the primacy of relationships, is it any wonder that we are beginning to reconfigure our ideas about management in relational terms?

In motivation theory, our attention is shifting from the enticement of external rewards to the intrinsic motivators that spring from the work itself. We are refocusing on the deep longings we have for community, meaning, dignity, and love in our organizational lives. We are beginning to look at the strong emotions that are part of being human, rather than segmenting ourselves (love is for home, discipline is for work) or believing that we can confine workers into narrow roles, as though they were cogs in the machinery of production. As we let go of the machine models of work, we begin to step back and see ourselves in new ways, to appreciate our wholeness, and to design organizations that honor and make use of the totality of who we are.

The impact of vision, values, and culture occupies a great deal of organizational attention. We see their effects on organizational vitality, even if we can't quite define why they are such potent forces. We now sense that some of the best ways to create continuity of behavior are through the use of forces that we can't really see. Many scientists now work with the concept of fields—invisible forces that structure space or behavior. I have come to understand organizational vision as a field—a force of unseen connections that influences employees' behavior—rather than as an evocative message about some desired future state. Because of field theory, I believe I can better explain why vision is so necessary, and this leads me to new activities to strengthen its influence.

Our concept of organizations is moving away from the mechanistic creations that flourished in the age of bureaucracy. We have begun to speak in earnest of more fluid, organic structures, even of boundaryless organizations. We are beginning to recognize organizations as systems, construing them as "learning organizations" and crediting them with some type of self-renewing capacity. These are our first, tentative forays into a new appreciation for organizations. My own experience suggests that we can forego the despair created by such common organizational events as change, chaos, information overload, and cyclical behaviors if we recognize that organizations are conscious entities, possessing many of the properties of living systems.

Some believe that there is a danger in playing with science and abstracting its metaphors because, after a certain amount of stretch, the metaphors lose their relationship to the tight scientific theories that gave rise to them. But others would argue that all of science is metaphor—a hopeful description of how to think of a reality we can never fully know. I share the sentiments of physicist Frank Oppenheimer who says: "If one has a new way of thinking, why not apply it wherever one's thought leads to? It is certainly entertaining to let oneself do so, but it is also often very illuminating and capable of leading to new and deep insights" (in Cole 1985, 2).

"One learns to hope that nature possesses an order that one may aspire to comprehend."

—C.N. Yang

Chapter 1

Discovering an Orderly World

It has taken us a long while to get here—a nine-mile hike up a gradual ascent over rocky paths. My horse, newly trained to pack equipment and still an amateur, has bumped against my back, bruised my heels, and finally, unavoidably, stepped on my toe, smashing it against the inside of my boot. But it's been worth it. Here are the American Rockies at their clichéd best. The stream where I sit soaking my feet glistens on for miles I can't see, into green grasses that bend to the wind. There are pine trees, mountains, hawks, and off at the far edge of the meadow a moose who sees us and moves to hide her great girth behind a four-inch-wide tree. The tree extends just to the edge of each eyeball. We laugh, but I suspect there's a lesson in it for all of us.

For months, I have been studying process structures—things that maintain form over time yet have no rigidity of structure. This stream that swirls around my feet is the most beautiful one I've encountered. Because it is vacation, I resist thinking too deeply about this stream; but as I relax into its flow, metaphors stir and gently whorl the surface.

Finally, I ask directly: What is it that streams can teach me about organizations? I am attracted to the diversity I see, to these swirling combinations of mud, silt, grass, water, rocks. This stream has an impressive ability to adapt, to shift the configurations, to let the power balance move, to create new structures. But driving this adaptability, making it all happen, I think, is the water's need to flow. Water answers to gravity, to downhill, to the call of ocean. The forms change, but the mission remains clear. Structures emerge, but only as temporary solutions that facilitate rather than interfere. There is none of the rigid reliance on single forms, on true answers, on past practices that I have learned in business. Streams have more than one response to rocks; otherwise, there'd be no Grand Canyon. Or else Grand Canyons everywhere. The Colorado realized that there were ways to get ahead other than by staying broad and expansive.

Organizations lack this kind of faith, faith that they can accomplish their purposes in various ways and that they do best when they focus on direction and vision, letting transient forms emerge and disappear. We seem fixated on structures; and we build them strong and complex because they must, we believe, hold back the dark forces that are out to destroy us. It's a hostile world out there, and organizations, or we who create them, survive only because we build crafty and smart—smart enough to defend ourselves from the natural forces of destruction. Streams have a different relationship with natural forces. With sparkling confidence they know that their intense yearning for ocean will be fulfilled, that nature creates not only the call, but the answer.

Many of the organizations I experience are impressive fortresses. The language of defense permeates them: in CYA memo-madness; in closely guarded personnel files; in activities defined as "campaigns," "skirmishes," "wars," "turf battles," and the ubiquitous phrases of sports that describe everything in terms of offense and defense. Some organizations defend themselves superbly even against their employees with regulations, guidelines, time clocks, and policies and procedures for every eventuality. One organization I worked in welcomed its new employees with a list of twenty-seven offenses for which they would be summarily fired—and the assurance that they could be fired for other reasons as well.

Some organizations have rigid chains of command to keep people from talking to anyone outside their department; and in most companies, protocols define who can be consulted, advised, or criticized. We are afraid of what would happen if we let these elements of the organization recombine, reconfigure, or speak truthfully to one another. We are afraid that things will fall apart.

This need to hold the world together, these experiences of fright and fragility, are so pervasive that I wondered about the phenomenon long before I came upon this teacher stream. Something that is everywhere must come to us from somewhere. But where? In twentieth-century Western thought, I believe the source is our fuzzy understanding of concepts that originated with seventeenth-century science. Three centuries ago, when the world was seen as an exquisite machine set in motion by God—a closed system with a watchmaker father who then left the shop—the concept of entropy entered our collective consciousness. Machines wear down; they eventually stop. In Yeats's phrase, "Things fall apart; the centre cannot hold, mere anarchy is loosed upon the world" (1956, 184-185). This is a universe, we feel, that cannot be trusted with growth, rejuvenation, process. If we want progress, then we must provide the energy, the momentum, to reverse decay. By sheer force of will, because we are the planet's consciousness, we will make the world hang together. We will resist death.

What a fearful posture this has been! Something Atlas only imagined, it has gone on so long. It is time to stop now. It is time to take the world off our shoulders, to lay it gently down and look to it for an easier way. It is not only streams that have something to teach us. Lessons are everywhere. But the question is key. If not with us, then where are the sources of order to be found?

I believe nature is abundantly littered with examples and lessons of order. Despite the experience of fluctuations and changes that disrupt our plans, the world is inherently orderly. And fluctuation and change are part of the very process by which order is created.

One of these examples is found in the concept of *autopoiesis* (from the Greek for self-production.) The spelling of this word always eludes me, although looking at it now, I realize it's just another example of "i" before "e." Its meaning, however, has been imprinted on my consciousness: "The characteristic of living systems to continuously renew themselves and to regulate this process in such a way that the integrity of their structure is maintained" (Jantsch 1980, 7). Autopoiesis—natural processes that support the quest for structure, process, renewal, integrity. This description is not limited to one type of organism—it describes life itself. Every living thing expends energy and will do whatever is needed to preserve itself, including changing. Every living thing exists in form and is recognizable as itself. But, as systems scientist Erich Jantsch writes, within that "globally stable state" a living system is constantly changing. It is "a never resting structure" that constantly seeks its own self-renewal (1980, 10).

Autopoietic structures are instructive in several ways. They illuminate an important paradox: Each structure has a unique identity, a clear boundary, yet it is merged with its environment. At any point in its evolution, the structure is noticeable as a separate event, yet its history is tied to the history of the larger environment and to other autopoietic structures (Briggs and Peat 1989, 154). What we observe, in ourselves as well as in all living entities, are boundaries that both preserve us from and connect us to the infinite complexity of the outside world. Autopoiesis, then, points to a different universe. Not the fragile, fragmented world we attempt to hold together, but a universe rich in processes that support growth and coherence, individuality and community.

Dissipative structures in chemistry also teach a paradoxical truth, that disorder can be the source of new order. Ilya Prigogine coined the term "dissipative structures" to focus attention on the inherent contradiction of the two descriptors (1980). Dissipation describes a loss, a process by which energy gradually ebbs away. Yet Prigogine discovered that such dissipative activity could play a constructive role in the creation of new structures. Dissipation didn't lead to the demise of a system. It was part of the process by which the system let go of its present form so that it could reemerge in a form better suited to the demands of the present environment.

Prigogine's work has helped explain a contradiction in Western science. If entropy is the rule, why does life flourish? Why is evolution in living systems related to progress and complexification, not to deterioration and disintegration?

In a dissipative structure, things in the environment that disturb the system's equilibrium play a crucial role in creating new forms of order. As the environment becomes more complex, generating new and different information, it provokes the system into a response. New information enters the system as a small fluctuation that varies from the norm. If the system pays attention to this fluctuation, the information grows in strength as it interacts with the system and is fed back on itself (a process of autocatalysis). Finally, the information grows to such a level of disturbance that the system can no longer ignore it. At this point, jarred by so much internal disturbance and far from equilibrium, the system in its current form falls apart. But this disintegration does not signal the death of the system. In most cases the system can reconfigure itself at a higher level of complexity, one better able to deal with the new environment.

Dissipative structures demonstrate that *disorder* can be a source of *order*, and that growth is found in disequilibrium, not in balance. The things we fear most in organizations—fluctuations, disturbances, imbalances—need not be signs of an impending disorder that will destroy us. Instead, fluctuations are the primary source of creativity. Scientists in this newly understood world describe the paths between disorder and order as "order out of chaos" or "order through fluctuation" (Prigogine 1984). These are new principles that highlight the dynamics between chaos and creativity, between disruptions and growth.

Quantum theory introduces yet another level of paradox into our search for order. At the quantum level we observe a world where change happens in jumps, beyond our powers of precise prediction. This world has also challenged our beliefs about objective measurement, for at the subatomic level the observer cannot observe anything without interfering or, more importantly, participating in its creation. The strange qualities of the quantum world, and especially its influence in shaking our beliefs in determinism, predictability, and control, don't seem to offer any hope for a more orderly universe. But our inability to predict individual occurrences at the quantum level is not a result of inherent disorder. Instead, the results we observe speak to a level of quantum interconnectedness, of a deep order that we are only beginning to sense. There is a constant weaving of relationships, of energies that merge and change, of constant ripples that occur within a seamless fabric. There is so much order that our attempts to separate out discrete moments create the appearance of disorder.

We have even found order in the event that epitomizes total disorder—chaos. Chaos theory has given us images of "strange attractors"—computer pictures of swirling motion that trace the evolution of a system. A system is defined as chaotic when it becomes impossible to know where it will be next. There is no predictability; the system never is in the same place twice. But as chaos theory shows, if we look at such a system long enough and with the perspective of time, it always demonstrates its inherent orderliness. The most chaotic of systems never goes beyond certain boundaries; it stays contained within a shape that we can recognize as the system's strange attractor (see illustrations, pages 86 and 124).

Throughout the universe, then, order exists within disorder and disorder within order. We have always thought that disorder was the absence of the true state of order, even as we constructed the word: *dis*-order. But is chaos an irregularity, or is order just a brief moment seized from disorder? Linear thinking demands that we see things as separate states: One needs to be normal, the other exceptional. Yet there is a way to see this ballet of chaos and order, of change and stability, as two complementary aspects in the process of growth, neither of which is primary. As one system's scientist described it:

> A system now appears as a set of coherent, evolving, interactive processes which temporarily manifest in globally stable structures that have nothing to do with the equilibrium and the solidity of technological structures. Caterpillar and butterfly, for example, are two temporarily stabilized structures in the coherent evolution of one and the same system. (Jantsch 1980, 6)

While we have lusted for order in organizations, we have failed to understand its true nature. We have seen order reflected in the structures we erect, whether they be bright mirror-glass buildings or plans started on paper napkins. These structures take so much time, creativity, and tinkering attention that it is hard not to want them to be permanent. It is hard to welcome disorder as a full partner in the search for order when we have expended such effort to bar it from the gates. I find myself challenged by this new landscape of openness, of structures that come and go, of bearings gained not from the rigid artifacts of organization charts and job descriptions, but from directions arising out of deep, natural processes of growth and self-renewal. This is not an easy land to inhabit, not an easy concept in which to place faith—except that we're already living with the evidence that supports it. And most of us have already experienced systems naturally recreating themselves as they play in their environment.

When I think about the team experiences I cherish most, I see such a pattern. In the interest of getting things done, our roles and tasks moved with such speed that the lines between structure and task blurred to nothing. When we speak of informal leadership, we describe a similar experience—the capacity of the organization to create the leadership that best suits its needs at the time. We may fail to honor these leaders more formally, trapped as we are in a hierarchical structure that is non-adaptive; but at the level of the living, where the people are, we know who the leader is and why he or she needs to be there. Max De Pree, former CEO of Herman Miller, calls this "roving leadership, the indispensable people in our lives who are there when we need them" (1989, 41-42). They emerge from the group, not by self-assertion, but because they make sense, given what the group needs to thrive and what individuals need to grow.

All this time, we have created trouble for ourselves in organizations by confusing control with order. This is no surprise, given that for most of its written history, management has been defined in terms of its control functions. Lenin spoke for many managers when he said: "Freedom is good, but control is better."

But our quest for control has been as destructive as was his.

If organizations are machines, control makes sense. If organizations are process structures, then seeking to impose control through permanent structure is suicide. If we believe that acting responsibly means exerting control by having our hands into everything, then we cannot hope for anything except what we already have—a treadmill of effort and life-destroying stress.

What if we could reframe the search? What if we stopped looking for control and began, in earnest, the search for order? Order we will find in many places we never thought to look before—all around us in nature's living, dynamic systems that are open to their environment. In fact, once we begin to look into nature with new eyes, the examples are overwhelming.

I looked again at the moose, staring intently into a narrow strip of tree bark. Our search for safety, our belief that we control our organizations by the structures we impose, is no less myopic. As long as we stare cross-eyed at that tree, we won't see all around us the innate processes of living systems that are there to create the order we crave.

Yet it is hard to step away from that tree. It is hard to open ourselves to a world of inherent orderliness. "In life, the issue is not control, but dynamic connectedness," Jantsch writes (1980, 196). I want to act from that knowledge. I want to move into a universe I trust so much that I give up playing God. I want to stop holding things together. I want to experience such safety that the concept of "allowing"—trusting that the appropriate forms can emerge—ceases to be scary. I want to surrender my care of the universe and become a participating member, with everyone I work with, in an organization that moves gracefully with its environment, trusting in the unfolding dance of order.

"For fragmentation is now very widespread, not only throughout society, but also in each individual; and this is leading to a kind of general confusion of the mind, which creates an endless series of problems and interferes with our clarity of perception so seriously as to prevent us from being able to solve most of them. . . . The notion that all these fragments are separately existent is evidently an illusion, and this illusion cannot do other than lead to endless conflict and confusion."

—David Bohm

CHAPTER 2

Newtonian Organizations in a Quantum Age

I sit in a room without windows, participating in a ritual etched into twentieth-century tribal memory. I have been here thousands of times before, literally. I am in a meeting, trying to solve a problem. Using whatever analytic tool somebody has just read about or been taught at their most recent training experience, we are trying to come to grips with a difficult situation. Perhaps it is poor employee morale or productivity. Or production schedules. Or the redesign of a function. The topic doesn't matter. What matters is how familiar and terrible our process is for coming to terms with the complaint.

The room is adrift in flip chart paper—clouds of lists, issues, schedules, plans, accountabilities—crudely taped to the wall. They crack and rustle, fall loose, and, finally, are pulled off the walls, tightly rolled, and transported to some innocent secretary, who will litter the floor around her desk so that, peering down from her keyboard, she can transcribe them to tidy sheets, which she will mail to us. They will appear on our desks days or weeks later, faint specters of commitments and plans, devoid of even the little energy and clarity that sent the original clouds—poof—up onto the wall. They will drift into our day planners and onto individual "to do" lists, lists already fogged with confusion and inertia. Whether they get "done" or not, they will not solve the problem.

I am weary of the lists we make, the time projections we spin out, the breaking apart and putting back together of problems. It does not work. The lists and charts we make do not capture experience. They only tell of our desire to control a reality that is slippery and evasive and perplexing beyond comprehension. Like bewildered shamans, we perform rituals passed down to us, hoping they will perform miracles. No new wisdom teacher has appeared to show us how to fit more comfortably into the universe. Our world grows more disturbing and mysterious, our failures to predict and control leer back at us from many places, yet to what else can we turn? If the world is not linear, then our approaches cannot work. But then, where are we?

The search for new shamans has begun in earnest. At the end of the twentieth century, our seventeenth-century organizations are crumbling. We have prided ourselves, in all these centuries since Newton and Descartes, on the triumphs of reason, on the absence of magic. Yet we, like the best magicians of old, have been hooked on prediction. For three centuries, we've been planning, predicting, analyzing the world. We've held onto an intense belief in cause and effect. We've raised planning to the highest of priestcrafts and imbued numbers with absolute power. We look to numbers to describe our economic health, our productivity, our physical well-being. We've developed graphs and charts and maps to take us into the future, revering them as ancient mariners did their chart books. Without them, we'd be lost, adrift among the dragons. We have been, after all, no more than sorcerers, the master magicians of the late twentieth century.

The universe that Sir Isaac Newton described was a seductive place. As the great clock ticked, we grew smart and designed the age of machines. As the pendulum swung with perfect periodicity, it prodded us on to new discoveries. As the Earth circled the sun (like clockwork), we grew assured of the role of determinism and prediction. We absorbed expectations of regularity into our very beings. And we organized work and knowledge to fit this universe.

It is interesting to note just how Newtonian most organizations are. The machine imagery of the spheres was captured by organizations in an emphasis on structure and parts. Responsibilities have been organized into functions. People have been organized into roles. Page after page of organizational charts depict the workings of the machine: the number of pieces, what fits where, who the big pieces are. Several years ago, we became concerned with the machine-based concept of organizational "fit," evidenced by its extensive literature. (William Bygrave, a physicist turned organizational theorist, comments on how many management strategy theorists either were engineers, or admired that profession, from Chandler to Porter. There has been a close connection, he writes, between their scientific training and their attempts to create a systematic, rational approach to business strategy [1989, 16].)

This reduction into parts and the proliferation of separations has characterized not just organizations, but everything in the world during the past three hundred years. Knowledge was broken into disciplines and subjects, engineering became a prized science, and people were fragmented—counseled to use different "parts" of themselves in different settings.

In organizations, we focused our attention on structure and organizational design, on gathering extensive numerical data, and on making decisions using sophisticated mathematical ratios. We've spent years moving pieces around, building elaborate models, contemplating more variables, creating more advanced forms of analysis. Until recently, we really believed that we could study the parts, no matter how many of them there were, to arrive at knowledge of the whole. We have reduced and described and separated things into cause and effect, and drawn the world in lines and boxes.

A world based on machine images is a world filled with boundaries. In a machine, every piece knows its place. Likewise, in Newtonian organizations, we've drawn boundaries everywhere. We've created roles and accountabilities, drawing lines of authority and limits to responsibilities. We have even drawn boundaries around the flow of experience, thus shaping the way we think about the world. For example, we are conditioned to think of reality in terms of variables. We study these variables for their allegedly dependent and independent properties, treating them as separate and well-bounded, even when we try to account for more of their interactions in multiple regression analyses. In business, information is portrayed in charts that chunk up the world. Pie charts tell us about market share, employee opinions, customer ratings. We have even come to think of power—an elusive, energetic force if ever there was one—as a measurable resource, defined by "our share of the pie."

These omnipresent boundaries create a strong sense of solidity, of structures that secure things, a kind of safety. Although we speak nowadays of boundaryless organizations, it is hard to picture life in such an organization, conditioned as we are to boundaries that protect and define.

Boundaries also create a strong sense of identity. They make it possible to know the difference between one thing and another. "The whole corpus of classical physics," writes Danah Zohar in *The Quantum Self*, "and the technology that rests on it is about the separateness of things, about constituent parts and how they influence each other across their separateness" (1990, 69). Classical physics studies a world of things and how connections work across the separations. In this world of things, there are well-defined edges; it is possible to tell where one stops and the other begins, to observe something without interfering with its identity or functioning. The "thing" view of the world, therefore, leads to a belief in scientific objectivity. And we prospered with this belief for many centuries, working well in a world of you-me, inside-outside, here-thereness.

A vast and complex machine had been entrusted to our care. We searched to know the mind of the clock maker, even as he receded deep into the distance. We made some assumptions about him (gender was never a question): He was infinitely rational, his works were totally predictable, and a few simple laws would reveal what made everything work. Reductionist thinking proliferated. God, it seemed, had been the primordial sponsor of KISS management—Keep It Simple, Stupid. "Chaos was merely complexity so great," Briggs and Peat comment, "that in practice scientists couldn't track it, but they were sure that in principle they might one day be able to do so. When that day came there would be no chaos, so to speak, only Newton's laws. It was a spellbinding idea" (1989, 22).

In physics, this search for simplicity led to work on a unified theory, more recently dubbed the "theory of everything" (see Davies and Brown 1988). In management, our desire for predictive principles has a less noteworthy history. We seem to have confused principles with a propensity for endless aphorisms and simplistic statements about what makes for a well-run organization.

It has not been easy living in this universe. A mechanical world feels distinctly anti-human. As Zohar eloquently describes it: "Classical physics transmuted the living cosmos of Greek and medieval times, a cosmos filled with purpose and intelligence and driven by the love of God for the benefit of humans, into a dead, clockwork machine. . . . Things moved because they were fixed and determined; cold silence pervaded the once-teeming heavens. Human beings and their struggles, the whole of consciousness, and life itself were irrelevant to the workings of the vast universal machine" (1990, 18).

From the scientists' perspective, there were still more consequences. Though they had engaged in a successful dialogue with nature, a major outcome of their work, as Prigogine and Stengers describe it, "was the discovery of a silent world. This is the paradox of classical science. It revealed to men a dead, passive nature, a nature that behaves as an automaton which, once programmed, continues to follow the rules inscribed in the program. In this sense, the dialogue with nature isolated man from nature instead of bringing him closer to it. . . . It seemed that science debased everything it touched" (1984, 6).

Loneliness pervaded not only our science, but whole cultures. In America, we raised individualism to its highest expression, each of us protecting our boundaries, asserting our rights, creating a culture that, Bellah et al. writes, "leaves the individual suspended in glorious, but terrifying, isolation" (1985, 6).

In science, the beginning of the twentieth century heralded the end of Newton's domination. Discoveries of a strange world at the subatomic level could not be explained by Newtonian laws, and the path was opened for new ways of comprehending the universe. Newtonian mechanics still apply to our world and still contribute greatly to scientific advances, but a new and different science is required now to explain many phenomena. Quantum mechanics, the most successful theory ever developed in physics, does not describe a clock-like universe. It tells of a very different world.

> Most of the other giant steps in our understanding of nature were really *evolutionary* in that they sprang from previously established foundations: facts were

> reorganized, or connected in new ways, or seen in a different context. Quantum theory, however, broke away completely from those foundations; it dove right off the end. It could not (cannot) adequately be described in metaphors borrowed from our previous view of reality because many of those metaphors no longer apply. But the net result has not been to obscure reality or make the nature of things more elusive and murky. On the contrary, most physicists would agree that what quantum theory has brought to science is exactly the opposite—concreteness and clarity." (Cole 1985, 106)

The quantum world is weird, even to scientists. Two of its most outstanding theoreticians made strong comments about this. Niels Bohr warns that, "Anyone who is not shocked by quantum theory has not understood it. "And Erwin Shroedinger, reacting to some of its puzzles, says, "I don't like it and I'm sorry I ever had anything to do with it" (in Gribbin 1984, 5; frontispiece).

But the quantum world is not just disquietingly fascinating; it is also relevant. As more and more of life at the sub-atomic level is revealed, I believe it will give us potent images that will enrich our lives at the macro level. It may also be true that quantum phenomena apply somewhat to us large-sized objects, literally, more than we had thought. Our brain cells "are sensitive enough to register the absorption of a single photon . . . and thus sensitive enough to be influenced by the whole panoply of odd, quantum-level behavior" writes Zohar (1990, 79). And Wolf notes that "Instead of finding quantum mechanics restricted to ever tinier corners of the universe, we physicists are finding its applicability ever increasing to larger and larger neighborhoods of time and space" (1981, xiv).

Because it is so strange, chroniclers of the quantum world reach for new metaphors. Zohar depicts it as "a vast porridge of being where nothing is fixed or measurable, . . . somewhat ghostly and just beyond our grasp" (1990, 27). Capra sees it as "dynamic patterns continually changing into one another—the continuous dance of energy" (1983, 91). Others say that it is a place where "everything is interconnected like a vast network of interference patterns" (in Lincoln 1985, 34). In 1930, astronomer James Jeans spoke of an image that, for me, aptly describes this new world: "The universe begins to look more like a great thought than like a great machine" (in Capra 1983, 86).

When the world ceased to be a machine, when we began to recognize its dynamic, living qualities, many familiar aspects of it disappeared. In the work of quantum theorists, "things" have disappeared. Although some scientists still conduct an elusive search for the basic building blocks of matter, other physicists have abandoned this as a final, futile quest of reductionism. They gave up searching for things finite and discrete because, as they experimented to find elementary particles, they found "things" that changed form and properties as they responded to one another, and to the scientist observing them. "In place of the tiny billiard balls moved around by contact forces," Zohar writes, "there are what amount to so many patterns of active relationship, electrons and photons, mesons and nucleons that tease us with their elusive double lives as they are now position, now momentum, now particles, now waves, now mass, now energy—and all in response to each other and to the environment" (1990, 98).

In the quantum world, relationships are not just interesting; to many physicists, they are *all* there is to reality. One physicist, Henry Stapp, describes elementary particles as, "in essence, a set of relationships that reach outward to other things" (in Capra 1983, 81). Particles come into being ephemerally, through interactions with other energy sources. We give names to each of these sources—physicists still identify neutrons, electrons, etc. —but they are "intermediate states in a network of interactions." Physicists can plot the probability and results of the interactions, as in Feynmann and S Matrix (scattering matrix) diagrams, but no particle can be drawn independent from the others. What is important in these diagrams is the overall process by which elements meet and change; analyzing them for more individual detail is simply not possible (Zukav 1979, 248-50).

In organizations, we are at the edge of this new world of relationships, hoping the new charts are true, still fearing if we follow them, that we will fall off into nothing. A mariner, perched high in the crow's nest, sometimes cries "Land ho" on faith. Knowing what to look for, knowing how land appears on the horizon, knowing how to tell clouds from land—still, sometimes, the call is an act of faith. Sighting a world of quantum organizations requires such faith. But as we become

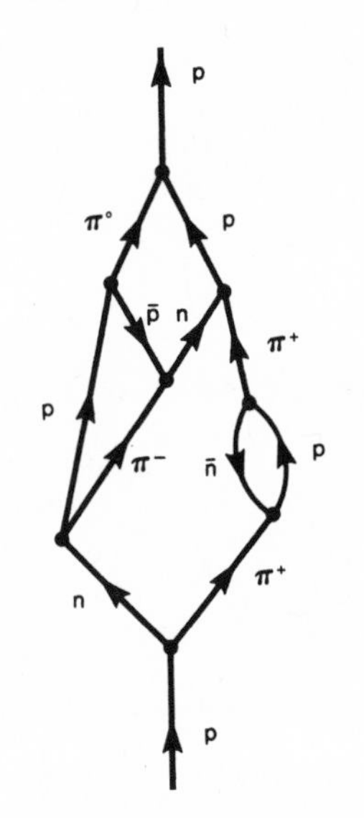

The multiple lives of a single proton that appear in a moment of time too brief for us to comprehend. In this reaction, charted by physicist Kenneth Ford, eleven particles make a brief appearance between the time the original proton transforms itself into a neutron and a pion and the time it becomes a single proton again (Zukav 1979, 237). This diagram illustrates the potentials that exist in each particle to exist as different combinations of other particles. These combinations are not random; the probability of their occurring can be calculated with accuracy, although, ultimately, which reactions occur is up to chance.

more familiar with the quantum world, a few of its organizational shapes emerge from the fog, their outlines just observable.

This world of relationships is rich and complex. Gregory Bateson (1980) speaks of "the pattern that connects," and urges that we stop teaching facts—the "things" of knowledge—and focus, instead, on relationships as the basis for all definitions. With relationships, we give up predictability for potentials. Several years ago, I read that elementary particles were "bundles of potentiality." I have begun to think of all of us this way, for surely we are as undefinable, unanalyzable, and bundled with potential as anything in the universe. None of us exists independent of our relationships with others. Different settings and people evoke some qualities from us and leave others dormant. In each of these relationships, we are different, new in some way.

If nothing exists independent of its relationship with something else, we can move away from our need to think of things as polar opposites. For years I had struggled conceptually with a question I thought important: In organizations, which is the more important influence on behavior—the system or the individual? The quantum world answered that question for me with an authoritative, "It depends." This is not an either/or question. There is no need to decide between the two. What is critical is the *relationship* created between the person and the setting. That relationship will always be different, will always evoke different potentialities. It all depends on the players and the moment.

Absolute prediction and uniformity are, therefore, impossible. While this may feel slightly unsettling, it certainly makes for a more interesting world. People go from being predictable to being surprising. Each of us is a different person in different places. This doesn't make us inauthentic, it merely makes us quantum.

Not only are we fuzzy, the whole universe is.

One source of universal fuzziness comes from the fact that elementary matter is inherently two-faced. It possesses two very different manifestations. Matter can be particles, localized points in space; or it can be waves, energy dispersed over a finite volume. Matter's total identity (known as a wave packet) includes potentialities for both forms—particles and waves. This is the Principle of Complementarity; and at heart, if I may give it a philosophical slant, it speaks of unity expressed as diversity.

However, these two complementary aspects of one existence cannot be studied simultaneously as a unified whole. Here, we are thwarted by another major principle of quantum physics, Heisenberg's Uncertainty Principle. We can measure position, and thus get a fix on the particle aspect; or we can study momentum, and observe the wave. But we can never measure both simultaneously. "While we can measure wave properties, or particle properties, the exact properties of the *duality* must always elude any measurement we might hope to make. The most we can hope to know about any given wave packet is a fuzzy reading of its position and an equally fuzzy reading of its momentum" (Zohar 1990, 27). It is this "vast porridge of being" that sucks into its quicksand all our hopes for a deterministic, quantifiable universe.

These two principles fundamentally change our relationship to measurement and observation. If quantum matter develops a relationship with the observer and changes to meet his or her expectation, then how can there be scientific objectivity? If the scientist structures experiments to study wave properties, matter behaves as a wave. If the experiment is to examine particles, matter shows up in particle form. The act of observation causes the potentiality of the wave packet to "collapse" into one aspect. Thus, one potential becomes enacted, while the others instantly disappear. (Physicists who postulate the Many Worlds [or parallel universes] theory say that no potentiality is lost. Each goes off, enacted, in its own world. Worlds upon worlds come into being and exist simultaneously [Wolf 1988].)

> Without perception, the universe continues . . . to generate an endless profusion of possibilities. The effect of perception is immediate and dramatic. All of the wave function representing the observed system collapses, except the one part, which actualizes into reality. No one knows what causes a particular possibility to actualize and the rest to vanish. The only law governing this is statistical. It is up to Chance." (Zukav 1979, 79)

No longer, in this relational universe, can we study anything as separate from ourselves. Our acts of observation are part of the process that brings forth the manifestation of what we are observing (see chapter 4). Particles remain as fuzzy bundles until they are observed. Only then do they become a thing. (At the moment the wave packet collapses, quantum phenomena give way and Newtonian physics reenters the picture.) John Archibald Wheeler, a noted physicist, states that the ultimate constituent of all there is in the universe is the "ethereal act of observer-participancy." The universe, he says, is a participative universe (in Zohar 1990, 45). We do not, as some have suggested, *create* reality, but we are essential to its coming forth. We *evoke a potential* that is already present. Because things cannot exist as observable phenomena without us in the quantum world, the ideal of scientific objectivity disappears.

Several years ago, organizational theorist Karl Weick called attention to *enactment* in organizations—how we participate in the creation of organizational realities. "The environment that the organization worries about is put there by the organization," he observed, adding that if we acknowledge the role we play in this creation, it changes the things we talk and argue about. If we create the environment, how can we argue about its objective features, or about what's true or false? Instead, Weick encouraged us to focus our concerns on issues of effectiveness, on questions of what happened, and what actions might have served us better. We could stop arguing about truth and get on with figuring what works best (1979, 152, 168-69).

Weick also suggested a new approach to organizational analysis. Acting should precede planning, he said, because it is only through action and implementation that we create the environment. Until we put the environment in place, how can we formulate our thoughts and plans? In strategic planning, we act as though we are responding to a demand from the environment; but, in fact, Weick argued, we *create* the environment through our own strong intentions. Strategies should be "just-in-time . . . , supported by more investment in general knowledge, a large skill repertoire, the ability to do a quick study, trust in intuitions, and sophistication in cutting losses" (1979, 223, 229). In other words, we should concentrate on creating organizational wave packets, resources that continue to expand in potential until needed.

Weick was describing a quantum world, although he used the term enactment. The environment remains uncreated until we interact with it; there is no describing it until we engage with it. Abstract planning divorced from action becomes a cerebral activity of conjuring up a world that does not exist.

So many of the things in organizations that we argue about, or spend time on, have been rooted in our belief in an objective reality. Something hard to define is out there, we believe, challenging our skills of analysis and perception. We have tried hard to become very smart so that we could accurately describe the environment. As we've grown more sophisticated, we have looked beyond simple cause and effect to multiple causes, for complex "fit" among many variables of strategy, structure, etc.

But this search for stable, well-defined targets has been, if we can admit it, a great cosmic joke. We thought we could pin reality down, get it in our sights, or maybe even line up our ducks; but how do you do that in this elusive world of potentials? We've been playing with "vast networks of interference patterns," with "the continuous dance of energy." The world is not a thing. It's a complex, never-ending, always changing tapestry.

To live in a quantum world, to weave here and there with ease and grace, we will need to change what we do. We will need to stop describing tasks and instead facilitate *process*. We will need to become savvy about how to build relationships, how to nurture growing, evolving things. All of us will need better skills in listening, communicating, and facilitating groups, because these are the talents that build strong relationships. It is well known that the era of the rugged individual has been replaced by the era of the team player. But this is only the beginning. The quantum world has demolished the concept of the unconnected individual. More and more relationships are in store for us, out there in the vast web of universal connections.

Even organizational power is a quantum event. One evening, I had a long, exploratory talk with a wise friend who told me that "power in organizations is the capacity generated by relationships." It is a real energy that can only come into existence through relationships. Ever since that conversation, I have changed what I pay attention to in an organization. Now I look carefully at how a workplace organizes its relationships; not its tasks, functions, and hierarchies, but the patterns of relationship and the capacities available to form them.

Because power is energy, it needs to flow through organizations; it cannot be confined to functions or levels. We have seen the positive results of this flowing organizational energy in our experiences with participative management and self-managed teams. What gives power its charge, positive or negative, is the quality of relationships. Those who relate through coercion, or from a disregard for the other person, create negative energy. Those who are open to others and who see others in their fullness create positive energy. Love in organizations, then, is the most potent source of power we have available. And all because we inhabit a quantum universe that knows nothing of itself, independent of its relationships.

This web of relationships is so dominant and yet so puzzling in quantum physics. Scientists have observed a level of connectedness among seemingly discrete parts that are widely separated in time and space. For many years, since 1930, a great debate raged among the premier physicists, especially between Niels Bohr and Albert Einstein. Could matter be affected by "non-local causes." Could it be changed by influences that travelled faster than the speed of light? Einstein was so repelled at the idea of a universe where cause could happen at a distance, that he, with two other physicists, designed a thought experiment (the EPR experiment) to disprove non-local causes. Thirty years later, with the debate still going, physicist John Bell constructed a mathematical proof to show that such a thing as "instantaneous action-at-a-distance" could occur in the universe. Finally, in 1982, Alain Aspect, a French physicist, conducted actual experiments proving that elementary particles are, indeed, affected by connections that exist unseen across time and space (Gribbin 1984, 227 ff).

Aspect's work was an extension of the original EPR experiment Einstein had postulated. In these experiments, two electrons, for example, are tested to determine if, once correlated (paired together), they can maintain their connection over distance. One way to correlate electrons is to pair their spins such that the sum of the spins is zero. Electrons spin along an axis of rotation that can be either "up" or "down." However, being quantum phenomena, the axes exist only as *potential* axes until we establish an axis to measure. Before measurement, we cannot describe a fixed spin to the electron; we only know that *when* we measure it, it will respond to the axis we choose to measure. In the two electrons we have paired together, if one is spinning up, the other will spin down, or if one is spinning right, the other will spin left.

During the experiment, the two electrons are separated. Theoretically, the distance can be across the universe. No matter the distance, at the moment one electron is measured for its spin—say that we choose a vertical axis—the second electron will instantaneously display a vertical, but opposite, spin also. How does this second electron, so far away, know which axis we chose to measure?

Formerly, we believed that nothing travels faster than the speed of light, yet here is evidence that disconfirms that. The explanation that physicists offer is that the two electrons are linked by non-visible connections; they are, in fact, an indivisible whole that cannot be broken into parts, even by spatial separation. When we attempt to measure them as discrete parts, we get stymied by their unseen fabric of connectedness.

In our day-to-day search for order and prediction, we are driven crazy because of non-local causality. In spite of the best of plans, we experience influences that we can't see or test, and strange occurrences pop up everywhere. We have broken things into parts and fragments for so long, and have believed that was the best way to understand them, that we are unequipped to see a different order that is there, moving the whole. British physicist David Bohm captures this dilemma when he says, "The notion that all these fragments are separately existent is evidently an illusion, and this illusion cannot do other than lead to endless conflict and confusion" (1983, 2).

At present, our most sophisticated way of acknowledging the world's complexity is to build elaborate systems and process maps, which are often influenced by a Newtonian quest for predictability. If we create the map to reveal all the variables, and expect that from such knowledge we will be able to manipulate the system for the outcomes we desire, we are thinking like Newton. What we hope for is not possible. There are no routes back to the safe harbor of prediction—no skilled mariners able to find their way across a deterministic ocean. The challenge for us is to see beyond the innumerable fragments to the whole, stepping back far enough to appreciate how things move and change as a coherent entity. We do, after all, live in a very fuzzy world, where boundaries have an elusive nature. The now-you-see-it, now-you-don't quality of these boundaries will continue to drive us crazy as long as we try to delineate them, or to decipher clear lines of cause and effect between well-bounded concepts.

There are no familiar ways to think about the levels of interconnectedness that seem to characterize the quantum universe. Instead of a lonely space, with isolated particles moving about, space appears filled with connections. This is why the metaphors turn to webs and weaving, or to the world as a great thought. Gravity is an everyday example of "action-at-a-distance," and scientists have created other "fields," unseen forces that structure space, to explain the connections they observe. (See chapter 3 for more about fields.) In an important way, though, even as we construct different field theories, we are still locked into a parts mentality, trying to explain how separate things join together. The more provocative view, expressed in Bohm's work, is that at a level we can't discern, there is an unbroken wholeness. If we could look beneath the surface, we would observe an "implicate order" out of which seemingly discrete events arise (1983).

I believe the evolving emphasis in our society to "think globally, act locally" expresses a quantum perception of reality. Acting locally is a sound strategy for changing large systems. Instead of trying to map an elaborate system, the advice is to work with the system that you know, one you can get your arms around. If we look at this strategy with Newtonian eyes, we would say that we are creating incremental change. Little by little, system by system, we develop enough momentum to affect the larger society.

A quantum view would explain the success of these efforts differently. Acting locally allows us to work with the movement and flow of simultaneous events within that small system. We are more likely to become synchronized with that system, and thus to have an impact. These changes in small places, however, create large-systems change, not because they build one upon the other, but because they share in the unbroken wholeness that has united them all along. Our activities in one part of the whole create non-local causes that emerge far from us. There is value in working with the system any place it manifests because unseen connections will create effects at a distance, in places we never thought. This model of change—of small starts, surprises, unseen connections, quantum leaps—matches our experience more closely than our favored models of incremental change.

The quantum leaps that we speak of so glibly also teach about quantum interconnectedness. Technically, these leaps are abrupt and discontinuous changes, where an electron jumps from one atomic orbit to another without passing through any intermediate stages. Exactly when the leap will occur is unpredictable; physicists can calculate the probability of a jump occurring, but not precisely when it will take place. What is at work here, though we cannot observe it, is a whole system creating the conditions that lead to the sudden jump. Because we don't know and can never know enough about the whole movement, we can never predict exactly how its influence will be manifest. This is hardly a comforting thought to those of us trying to manage organizations, yet quantum leaps reflect our experience of organizational change with more accuracy than we commonly acknowledge.

My growing sensibility of a quantum universe has affected my organizational life in several ways. First, I try hard to discipline myself to remain aware of the whole and to resist my well-trained desire to analyze the parts to death. I look now for patterns of movement over time and focus on qualities like rhythm, flow, direction, and shape. Second, I know I am wasting time whenever I draw straight arrows between two variables in a cause and effect diagram, or position things as polarities, or create elaborate plans and time lines. Third, I no longer argue with anyone about what is real. Fourth, the time I formerly spent on detailed planning and analysis I now use to look at the structures that might facilitate relationships. I have come to expect that something useful occurs if I link up people, units, or tasks, even though I cannot determine precise outcomes. And last, I realize more and more that the universe will not cooperate with my desires for determinism.

Those who have used music metaphors in describing leadership, particularly jazz metaphors, are on a quantum track. Improvisation is the saving skill. As leaders, we play a crucial role in selecting the melody, setting the tempo, establishing the key, and inviting the players. But that is all we can do. The music comes from something we cannot direct, from a unified whole created among the players—a relational holism that transcends separateness. In the end, when it works, we sit back, amazed and grateful.

As I was writing this chapter, I received a call from a client who was deep into a project and very frustrated. His organization had collected data, defined five key problem areas, and created task forces to solve each of those issues. Yet the managers were having problems coordinating the task forces. The longer the task forces studied the issues, the more they were seeing the problems as interrelated. Threads of interconnections were everywhere, yet the five groups were still acting autonomously from one another. The result was fatigue and impatience. People simply wanted to get on with implementing something; anything would be a relief after so many deadening meetings and detailed plans.

As I listened, I experienced my classic case of "Newtonian despair." I knew what he was feeling, I knew where things were headed, and I could not be very helpful. We talked for some time about bringing the whole system together to push for a deeper level of analysis, but it didn't satisfy my need for a richer vision of what to do.

I felt as Heisenberg must have, when he walked those streets at dawn, begging for new insights into the universe. I, too, can feel the ground shaking. I hear its deep rumblings. Any moment now, the earth will crack open and I will stare into its dark center. Into that smoking caldera, I will throw most of what I have treasured, most of the techniques and tools that have made me feel competent. I cannot do that yet; I cannot just heave everything I know into the abyss. But I know it is coming. And when it comes, when I have made my sacrificial offerings to the gods of understanding, then the ruptures will cease. Healing waters will cover the land, giving birth to new life, burying forever the ancient, rusting machines of our past understandings. And on these waters I will set sail to places I only now imagine. There I will be blessed with new visions and new magic. I will feel once again like a creative contributor to this mysterious world. But for now, I wait. An act of faith. Land ho.

"Although we know a great deal about the way fields affect the world as we perceive it, the truth is no one really knows what a field is. The closest we can come to describing what they are is to say that they are spatial structures in the fabric of space itself."

—Michael Talbot

CHAPTER 3

Space Is Not Empty: Invisible Fields that Shape Behavior

In Utah, the sky is everywhere—blue, open, insistent on attention. It soars over mountains and sweeps into long valleys, showing off its crystal clarity. At night, it is even more an exhibitionist. A friend, after a long flight from Hartford, sat rocking on our lawn swing far past midnight, tired, yawning, but unable to move. The stars would not let her go. For me, moving here—and living with these stars and sky—has been an experience in space. I have felt myself expanding into this vastness, felt my boundaries open, my vision lift, my internal defenses dissolve. With so much space, there is no place to go but out.

Space is the basic ingredient of the universe; there is more of it than anything else. Even at the microscopic level of atoms, where we would expect things to be small and compact, there is mostly space. Within atoms, subatomic particles are separated by vast distances, so much so that an atom is 99. 999 percent empty. Everything we touch, including our bodies, is composed of these empty atoms. We are far more porous than our dense bodies indicate. In fact, we are as void, proportionately, as intergalactic space (Chopra 1989, 96).

In Newton's universe, the emptiness of space created a sense of unspeakable loneliness. Matter, small and isolated, moved bravely through the void, intent on its solo track, meeting rarely, traveling always across wide gulfs that stretched on into infinity. This lonely universe has, for a long time, affected our self-expression in all ways, from existentialist philosophies where we have felt the need to create the meaning of our lives in isolation from any greater source, to the heroic individuals of American history, lonely champions (both Western and corporate) who succeeded in spite of great odds. It was difficult to effect change in such an open, lonely world. It required generating energy of sufficient strength to surge through space, enduring long enough to reach another object and cause it to respond. Newton's world of cause and effect, of force acting upon force, required such efforts—great expenditures of personal energy to get someone else moving, vast regions of space to traverse to get something done. Not only did it feel lonely, it was exhausting.

Something strange has happened to space in the quantum world. No longer is there a lonely void. Space everywhere is now thought to be filled with fields, invisible, non-material structures that are the basic substance of the universe. We cannot see these fields, but we do observe their effects. They have become a useful construct for explaining action-at-a-distance, for helping us understand why change occurs without the direct exertion of material "shoving" across space.

In scientific thought, field theory developed in several areas years before quantum physics as an attempt to explain action-at-a-distance. (The word *field* was taken from the name given to the background on heraldic shields.) Newton introduced the first field, gravitation. In his model, gravity originated from a center of force, such as the earth, and spread out from there into space. Imaginary lines of force filled space, attracting objects toward the earth. In Newton's model of gravitational pull, a force emanated from one source, acting on another.

Einstein developed a different view of the gravitational field. In relativity theory, gravity acts to structure space. The reason objects are drawn to earth is because space-time curves in response to matter. Rather than a force, gravity is understood as a medium, an invisible geometry of space.

In our day-to-day lives, we have direct experience with fields besides gravity. Just place iron filings near a magnet. The specific patterns that form around the magnet are due to the invisible magnetic field. We also experience the effects of fields every time we turn on a light or plug in an appliance. Our modern electrical generating stations spin huge magnets, creating magnetic fields that then create electrical fields, which send out currents of electrons.

Fields are conceived of in many different ways, depending on the theory. The gravitational field is conceived as a curved structure in space-time; electromagnetic fields create disturbances that manifest themselves as electromagnetic radiations; quantum fields, perhaps a different field for each particle, produce particles when two fields intersect. But in each of these theories, fields are unseen structures, occupying space and becoming known to us through their effects (see Sheldrake 1981, 60; Wilczek and Devine 1988, 155-64; Zukav 1979, 199-200).

Early advances in field theory came about because nineteenth-century scientists such as Michael Faraday and James Maxwell chose to concentrate not on specific particles, but on space. Intuitively, they sensed that space was not empty but instead was, in a modern physicist's phrase, "a cornucopia of invisible but powerfully effective structure" (Wilczek and Devine 1988, 156). Faraday and Maxwell made a conscious shift in vision, as we do when we look from close to distant objects; and in that shift, they led the way into a universe of busy, bustling space. It was an important shift in focus—to look behind the small, discrete, visible structures to an invisible world filled with mediums of connections.

Frank Wilczek and Betsy Devine, he a physicist, she an engineer turned writer, created an effective image for thinking about these invisible fields that exert visible influence. If we were to observe fish, unaware of the medium of water in which they swim, we would probably look for explanations of their movements in terms of one fish influencing another. If one fish swam by, and we observed the second fish swerving a little, we might think that the first was exerting a force on the second. But if we observed all the fish deflecting in a regular pattern, we might begin to suspect that some other medium was influencing their movements. We could test for this medium, even if it were still invisible to us, by creating disturbances in it and noting the reactions of the fish. The space that is everywhere, from atoms to the sky, is more like this ocean, filled with fields that exert influence and bring matter into form.

The quantum world is teasing and enticing in many ways. Fields fit right in. They are, as biologist Rupert Sheldrake describes them, "invisible, intangible, inaudible, tasteless and odorless" (1981, 72). They are often unapproachable through our five senses; yet in quantum theory, they are as real as particles. They are, says writer Gary Zukav, the substance of the universe. The things we see or observe in experiments, the physical manifestations of matter as particles, are a secondary effect of fields. Particles may come into existence, often temporarily and briefly, when two fields intersect. At the point of intersection, where their energies meet, particles appear. The fact that particles appear and disappear like quick-change artists is a result of continual interactions between different fields. Although we have thought of particles as the basic building blocks of matter, in fact they are oftentimes transitory, just brief moments of meeting recorded as observable matter. This leads to a puzzling situation. Physical reality is not only material. Fields are considered real, but they are non-material.

This paradox pushes us into important new territory, shoving us farther away from our "thing" thinking or away from a universe of parts linked tenuously by energetic forces. Fields encourage us to think of a universe that more closely resembles an ocean, filled with interpenetrating influences and invisible structures that connect. This is a much richer portrait of the universe; in the field world, there are potentials for action everywhere, anywhere two fields meet. "The Newtonian picture of a world populated by many, many particles, each with an independent existence, has been replaced by the field picture of a world permeated with a few active media. We live amid many interpenetrating fields—each filling space. The laws of motion, in field language, are rules for flows in this ocean. And the rules of transformation are, in this picture, telling us what . . . reactions occur among the components of the universal ocean" (Wilczek and Devine 1988, 163).

Sheldrake has created an intriguing concept of fields in biology. He has postulated the existence of morphogenic fields that govern the behavior of species. This type of field possesses very little energy, but it is able to take energy from another source and shape it. The field acts as a geometrical influence, shaping behavior. Morphogenic fields are built up through the accumulated behaviors of species' members (Sheldrake 1981, 60). After part of the species has learned a behavior, such as bicycle riding, others will find it easier to learn that skill. The form resides in the morphogenic field, and when individual energy combines with it, it patterns behavior without the need for laborious learning of the skill. These fields, saysBohm, provide "a quality of form that can be taken up by the energy of the receiver" (in Talbot 1987, 68; see also Sheldrake 1988).

Field images, when applied to organizations and employees, become quite provocative. We can imagine organizational space in terms of fields, with employees as waves of energy, spreading out in regions of the organization, growing in potential. How do we tap into that energy? How do we turn the employees' energy into behavior for the organization, into something observable and probable? The answer, in a field-filled world, is that employees meet up with other fields. Whether it's a field of energy, or a quality of form, they will have to interact with it to have their behavior made manifest. Space is not empty. Unseen energies influence how we manifest. The question becomes: What are the fields in organizations?

Before I propose an answer, let me note that we already have been developing a new awareness of organizational space. I listened recently to a senior insurance company executive describe his newly reorganized company. With advanced levels of electronic interconnectivity, he said, "space will be filled with our invisible, electronic networks, reaching everywhere on the globe." The computer world has also created new images with *Cyberspace*, a term used to describe air filled with information that we retrieve electronically. Electronically-generated information, invisible but essential, floating along the airwaves, retrievable from who-knows-where, is making space more of an active player in our organizations. But it is time to think about the field qualities of organizational space as well.

As a generation of managers, we have been focused on many of the ethereal qualities of organizations—culture, values, vision, ethics. Each of these words describes a *quality* of organizational life that we can observe in our experience, yet find elusive to pin down in specifics. Recently, while doing work on customer service for a retail chain, I asked employees to visit several stores. After spending time in many stores, we all compared notes. To a person, we agreed that we could "feel" good customer service just by walking into the store. We tried to get more specific by looking at visual cues, merchandise layouts, facial expressions—but none of that could explain the sure sense we had when we walked into the store that we would be treated well. Something else was going on. Something else was in the air. We could feel it, we just couldn't describe *why* we felt it.

It seems to me that field theory provides a useful explanation to this and many other organizational mysteries. At one level, thinking about organizational fields is metaphoric, an interesting concept to play with. But the longer I have thought about it, the more I am willing to believe that there are literal fields in organizations. I can imagine an invisible customer service field filling the spaces of those stores we visited, helping to structure employees' activities, and generating service behaviors whenever the energy of an employee intersected with that field. Of course, the field didn't just drift into the store; the cosmos didn't create a customer service field. In each of those stores there was a manager who, together with employees, took time to fill the store space with clear messages about how he or she wanted customers to be served. Clarity about service filled every nook and cranny. With such a powerful structuring field, certain types of individual behaviors and events were guaranteed.

This is not a fantasy image. Field theory can educate us in several ways about how to manage the more amorphous sides of organizations. For example, *vision*—the need for organizational clarity about purpose and direction—is a wonderful candidate for field theory. In linear fashion, we have most often conceived of vision as thinking into the future, creating a *destination* for the organization. We have believed that the clearer the image of the destination, the more force the future would exert on the present, pulling us into that desired future state. It's a very strong Newtonian image, much like the old view of gravity. But what if we changed the science and looked at vision as a field? What if we saw a field of vision that needed to permeate organizational space, rather than viewing vision as a linear destination?

If vision *is* a field, think about what we could do differently to create one. We would do our best to get it permeating through the entire organization so that we could take advantage of its formative properties. All employees, in any part of the company, who bumped up against that field would be influenced by it. Their behavior could be shaped as a result of "field meetings," where their energy would link with the field's form to create behavior congruent with the organization's goals. In the absence of that field, in areas of the organization that hadn't been reached, we could hold no expectation of desired behaviors. If the field hadn't extended into that space, there would be nothing there to help behaviors materialize, no invisible geometry working on our behalf.

Several years ago, a garbage can metaphor was introduced into our thinking about organizations. It created a provocative view of organizational "space" as a continual mixture of people, solutions, choices, and problems flowing around, and every so often coinciding and creating a decision at that juncture. "An organization is a collection of choices looking for problems, issues and feelings looking for decision situations in which they might be aired, solutions looking for issues to which they might be the answer, and decision makers looking for work" (Cohen, March, and Olsen 1974).

This is a potent view of a Newtonian organization, with discrete pieces wandering about, colliding or avoiding collision, veering off in unexpected directions—organizational anarchy relieved by occasional moments of accidental coherence. This metaphor has lasted for many years because it matches many of our experiences of messy and irrational organizational forces. The task of creating order in garbage cans, of imposing structure and clarity, is overwhelming.

But with a quantum world view, there are new possibilities for order. Organizational space can be filled with the invisible geometry of fields. Fields, being everywhere at once, can connect discrete and distant actions. Fields, because they can influence behavior, can cohere and organize separate events.

In many ways, we already know what powerful organizers fields can be. We have moved deeper into a field view of reality by our recent focus on culture, vision, and values as the means for managing organizations. We know that this works, even when we don't know how to do it well. Robert Haas, CEO of Levi Strauss & Co. , calls this phenomenon "conceptual controls. . . . It's the ideas of a business that are controlling, not some manager with authority" (in Howard 1990, 134). If we think of ideas as fields, I believe we have a better metaphor for understanding why concepts control as well as they do. But it changes the nature of our attention.

In a field view of organizations, clarity about values or vision is important, but it's only half the task. Creating the field through the dissemination of those ideas is essential. The field must reach all corners of the organization, involve everyone, and be available everywhere. Vision statements move off the walls and into the corridors, seeking out every employee, every recess of the organization. In the past, we may have thought of ourselves as skilled crafters of organizations, assembling the pieces of an organization, exerting our energy on the painstaking creation of links between all those parts. Now, we need to imagine ourselves as broadcasters, tall radio beacons of information, pulsing out messages everywhere. We need all of us out there, stating, clarifying, discussing, modeling, filling all of space with the messages we care about. If we do that, fields develop—and with them, their wondrous capacity to bring energy into form.

Field creation is not just a task for senior managers. Every employee has energy to contribute; in a field-filled space, there are no unimportant players. Sheldrake's morphogenic fields grow and develop form because of what is occurring at the level of the individual who is acquiring new skills and knowledge. These fields change their content and shape because of individual activity. This is similar to the insights of organizational consultant Peter Senge, who believes that an organization's vision grows as a by-product of individual visions, a by-product of ongoing conversations (1990, 212).

In the natural world, fields once formed may continue to propagate, continuing to exist even when their progenitors have disappeared. This is a comforting and cautioning trait that must influence our thinking about organizational fields. For once we create them, once we invest resources in putting them out there, they will sustain themselves, perhaps exerting more control than we had planned.

We need, therefore, to be very serious about this work of field creation, because fields give form to our words. If we have not bothered to create a field of vision that is coherent and sincere, people will encounter other fields, the ones we have created unintentionally or casually. It is important to remember that space is never empty. If we don't fill it with coherent messages, if we say one thing but do another, then we create dissonance in the very *space* of the organization. As employees bump up against contradicting fields, their behavior mirrors those contradictions. We end up with what is common to many organizations, a jumble of behaviors and people going off in different directions, with no clear or identifiable pattern. Without a coherent, omnipresent field, we cannot expect coherent organizational behavior. What we lose when we fail to create consistent messages, when we fail to "walk our talk," is not just personal integrity. We lose the partnership of a field-rich space that can help bring form and order to the organization.

There is an irony here. Those who try to convince us to manage from values or vision, rather than from traditional authority, usually scare us. Their organizations seem devoid of the management controls that ensure order. Values, vision, ethics—these are too soft, many feel, too translucent to serve as management tools. How can they create the kind of order we crave in the face of chaos? Newton's world justified those fears because it was a world of pieces spinning off in all directions. But if we look past Newton, if we change our field of vision, we see a world of greater, more subtle forms of order.

What if we slip out quietly along the curvature of space, out into its far reaches? What if, once there, we adjust our eyes magically to the invisible world? There we will see a plenitude of structure—potential structure, emerging structure—and we will stop doubting. We once were made to feel secure by things visible, by structures we could see. Now it is time to embrace the invisible. In a world where matter can be immaterial, where the substance of everything is something we can't see, why not ally ourselves with fields? For such a little act of faith, space awaits, filled with possibilities.

"If the world exists and is not objectively
solid and preexisting before I come on the scene,
then what is it? The best answer seems to be
that the world is only a potential and
not present without me or you to observe it.
It is, in essence, a ghost world that pops
into solid existence each time one of us observes it.
All of the world's many events are potentially
present, able to be but not actually seen
or felt until one of us sees or feels."

—Fred Allen Wolf

CHAPTER 4

The Participative Nature of the Universe

Schroedinger's cat is a classic thought problem in quantum physics. Physicist Erwin Schroedinger constructed the problem in 1935 to illustrate that in the quantum world nothing is real. We cannot know anything about what is happening to something if we are not looking at it, and, stranger yet, nothing *has* happened to it until we observe it. Central to the quantum world, Zohar wrote, is the idea that "unobserved quantum phenomena are radically different from observed ones" (1990, 41).

The problem of the cat has not yet been resolved, but it is constructed as follows: A live cat is placed in a box. The box has solid walls, so no one outside the box can see into it. This is a crucial factor, since the problem centers on the role of the observer in evoking reality. A device will trigger the release of either poison or food; the probability of either occurrence is 50/50. Time passes. The trigger goes off. The cat meets its fate.

Or does it? Just as an electron is *both* a wave and a particle until our observation causes it to collapse as *either* a particle or wave, Schroedinger argues that the cat is both alive *and* dead until the moment we observe it. Inside the box, unobserved, the cat exists only as a probability wave. It is possible to calculate mathematically (as a Schroedinger wave function) all of the cat's possible states. But it is impossible to say that the cat is living or dead until we observe it. It is the act of observation that determines the collapse of the cat's wave function and makes it either dead or alive. Before we peer in, the cat exists as probabilities. Our nosiness determines its fate.

I have *never* understood the quantum logic of Schroedinger's cat, but I have let the problem ramble aimlessly in my consciousness, content to not engage with its counterintuitive nature. Yet just like a wave function, the possibilities of this idea grew unobserved until one day, in true quantum fashion, they "popped" and I had a moment of concrete recognition. I realized I had been living in a Schroedinger's cat world in every organization I had ever been in. Each of these organizations had myriad boxes, drawn in endless renderings of organizational charts. Within each of those boxes lay a "cat," a human being, laden with potential, whose fate was determined, always and irrevocably, by the act of observation.

It is common to speak of self-fulfilling prophecies and the impact these have on the people we manage. If a manager is told that a new trainee is particularly gifted, that manager will see genius emerging from the trainee's mouth even in obscure statements. But if the manager is told that his or her new hire is a bit slow on the uptake, the manager will interpret a brilliant idea as a sure sign of sloppy thinking or obfuscation. From studies on the impact of opportunity in organizations (Kanter 1977), we know that the "anointed" in organizations, those high flyers who move quickly through the ranks, are given life through our desire to observe them as winners. We endow their ideas and words with more credibility. We entrust them with more resources and better assignments. We have already decided that they will succeed, and so we continually observe them with the expectation that they will confirm our beliefs.

Others in organizations go unobserved, irrevocably invisible, bundles of potential that no one bothers to look at. Or they receive summary glances, are observed to be "dead," and are thereafter locked into jobs that provide them with no opportunity to display their many potentials. In the quantum world, what you see is what you get. In human organizations, we play with Schroedinger's cat daily, determining the fate of all of us—our quality of aliveness or deadness—by the way we decide to observe one another. So it is not only quantum physicists who have to deal with the enigmas of observation. The observation problem is as real a dilemma for us as it is for them.

In quantum physics, the observation problem has led scientists to various schools of thought, each focused on the role played by consciousness. Is it consciousness that evokes the world? Is there any such thing as reality independent of our acts of observation? These questions have been asked not because of the physicists' interest in philosophy, but because the issues emerge in actual quantum experiments. The most frequently explained experiment that illustrates, among other things, the role of consciousness in quantum work is the double-slit experiment.

Most simply, this experiment involves electrons (or other elementary particles) that must pass through one of two openings (slits) in a surface. After passing through one of these slits, each electron lands on a second surface, where its landing is recorded. A single electron will only pass through one of the openings, but its behavior will be affected by whether one or both slits is open at the time it passes through either one of them.

The electron, like all quantum entities, has two identities, that of a wave, and that of a particle. If both slits are open, the electron acts as a wave, creating a pattern on the recording screen typical of the diffusion caused by a wave.

If only one slit is open, the resulting pattern is that associated with separate, particle behavior.

On its way through one slit, the electron acts in a way that indicates it "knows" whether or not the second hole is open. It knows what the scientist is testing for, and adjusts its behavior accordingly. If the observer tries to "fool" the subject by opening and shutting slits as the electron approaches the wall, the electron behaves in the manner appropriate for the state of the holes *at the moment* it passes through one. (For a detailed explanation of this experiment, see Gribbin 1984, 169-174.) The electron also knows if the observer is watching. If the recording apparatus is not on, the electron behaves differently than if it is being recorded. When the electron is not being observed, it exists only as a wave of probabilities; unless someone is watching, "nature herself does not know which hole the electron is going through" (Gribbin 1984, 171).

Physicist Richard Feynmann calls the double-slit experiment the basic element of quantum theory. Because nothing in it can be explained by classical physics, he dubs these experiments "the only mystery," that which contains all of "the basic peculiarities of quantum mechanics" (in Gribbin 1984, 164). As non-physicists, we may think we have an easier time with the mysteries of such things as observation and the role of the observer, but it seems to me we would do well to linger longer with these quandaries, to explore how our perceptions of people and events shape the reality we then end up struggling with so much.

Schroedinger's cat and the problem of observation pad quietly around our organizations in many forms. Fred Wolf, a physicist and translator into lay language of quantum physics, says that "knowing is disrupting." Every time we go to measure something, we interfere. A quantum wave function builds and builds in possibilities until the moment of measurement, when its future collapses into only one aspect. Which aspect of that wave function comes forth is largely determined by *what* we decide to measure.

The physicist John Archibald Wheeler has been an eloquent proponent of the participative universe, a place where the act of looking for certain information evokes the information we go looking for—and eliminates our simultaneous opportunity to observe other information. For Wheeler, the whole universe is a participatory process, where we create not only the present with our observations, but the past as well. It is the existence of observers who notice what is going on that imparts reality to the origin of everything (Gribbin 1984, 212). When we choose to experiment for one aspect, we lose our ability to see any others. Every act of measurement loses more information than it obtains, closing the box irretrievably and forever on other possibilities.

The difficulties of measurement raised by quantum sensibilities are not trivial issues for managers. We are addicted to numbers, taking frequent pulses of our organizations in surveys, monthly progress checks, quarterly reports, yearly evaluations. It is difficult to develop a new sensitivity to the fact that no form of measurement is neutral. Physicists call this awareness *contextualism*, a sensitivity to the interdependency between how things appear and the environment which causes them to appear. Contextualism raises some very important questions. How can we trust that we get the information we need to make intelligent decisions? How can we know what is the right information to look for? How can we remain sensitive to and retrieve the information we lost when we went looking for the information we got?

We don't often allow these questions to surface in organizations. We tend to focus on a few key indicators, or the words of those we trust. We worry more about the accuracy of the information we have and how best to analyze it, and seem less concerned about the information we lost by collecting the information now in front of us. We may look to new sources or different variables for missing data, but even then, we seem to believe that the data exists "out there" and that we just have to find the appropriate lens to focus it. We still believe in objectivity, in hard data, in firm numbers. We have avoided the murky, fuzzy world of non-objectivity that contextualism brings to the surface.

Yet how can we exist without objective information? How can we develop the information we need to do business, if the world is truly non-objective? The answer, I believe, is found in the participative nature of the universe. Participation, seriously done, is a way out from the uncertainties and ghostly qualities of this nonobjective world we live in. We need a broad distribution of information, viewpoints, and interpretations if we are to make sense of the world.

Let me develop a quantum interpretation as to why participation is such an effective organizational strategy. In the traditional model, we leave the interpretation of information to senior or expert people. Although they may be aware, to some extent, that they are interpreting the data, choosing some aspects of it, and ignoring others, few have been aware of how much potential data they lose through acts of observation. A few people, charged with interpreting the data, are, in fact, observing only very few of the potentialities contained within that data.

Think of organizational data for a metaphoric moment as a wave function, moving through space, developing more and more potential explanations. If this wave of potentialities meets up with only one observer, it will collapse into one interpretation, responding to the expectations of that particular observer. All other potentialities disappear from view and are lost by that act of observation. These data, already severely limited in potential meaning by the process of observation, are then assembled and passed down to others in the organization. Most often they are presented as objective, which they are not, and complete, which is an impossibility if we recall all those lost potentialities.

Consider how different it is, in quantum terms, when the wave of information spreads out broadly everywhere in the organization. Instead of "collapsing" into just a few interpretations, many moments of meeting—hundreds, even thousands of them—will occur. At each of those intersections between an observer and the data, an interpretation will appear, one that is specific to that act of observation. Instead of losing so many of the potentialities contained within the data wave, the multiplicity of interactions can elicit many of those potentials, giving a genuine richness to the data that is lost when we restrict information access to only a few people. An organization swimming in many interpretations can then discuss, combine, and build on them. The outcome of such a process has to be a much more diverse and richer sense of what is going on and what needs to be done.

It would seem that the more participants we engage in this participative universe, the more we can access its potentials and the wiser we can become. To banish the ghosts in this ghostly universe, we need a different pattern—one in which more and more of us engage freely, evoking multiple meanings through our powers of observation. "Whatever we call reality," Prigogine and Stengers advise, "it is revealed to us only through an active construction in which we participate" (1984, 293).

Certainly the most exciting and richly textured organizational events I have participated in recently are "Futures Search Conferences," where fifty to seventy people, drawn from all parts of the organization and from external constituent groups, work intensely together to create shared visions of the organization's past, present, and future. The richness of the interpretations and the multi-layered complexity of the future scenarios that are created have convinced me of the powers of observation and perception that participation brings forth. In these conferences, wave functions collapse into all sorts of strange and powerful interpretations because the whole system is in the room, generating information, thinking about itself and what it wants to be (see Weisbord, forthcoming).

The participative universe we inhabit has also expanded my understanding of the importance of "ownership," a concept whose definition has been changing. We now use the term not only to describe stockholder investments, but to talk of the emotional investment we want employees to have in their work. Ownership describes personal links to the organization, the charged, emotion-driven *feeling* that can inspire people. A tried and true maxim of my field of organizational behavior (O.B.) is that "people support what they create." Though I have preached, like every O.B. consultant before me, the values of psychological ownership, I now see that the quantum universe supports this concept even more strongly and explains *how* it creates real and tangible sources of energy.

We know that the best way to build ownership is to give over the creation process to those who will be charged with its implementation. We are never successful if we merely present a plan in finished form to employees. It doesn't work to just ask people to sign on when they haven't been involved in the design process, when they haven't experienced the plan as a living, breathing thing.

This is where the observation phenomenon of quantum physics has something to teach us. In quantum logic, it is impossible to expect any plan or idea to be real to employees if they do not have the opportunity to personally interact with it. Reality emerges from our process of observation, from decisions we the observers make about what we will see. It does not exist independent of those activities. Therefore, we cannot *talk* people into reality because there truly *is* no reality to describe if they haven't been there. People can only become aware of the reality of the plan by interacting with it, by creating different possibilities through their personal processes of observation.

Think about what happens in your experience when you want to get a plan accepted. I see it frequently in meetings where a plan is being proposed. Even if the plan is excellent, it will be a long meeting in which the plan will be dissected, criticized, thrown out, brought back, and finally, almost always, approved in its initial form with only a few slight modifications. All of those participants, like the best scientists, need to observe the plan in detail, exploring its edges, searching out its interior, playing with its potentialities. At each point of observation, ideas and energy and possibilities are brought forth and made real. Alternate views of reality are evoked that are felt, evaluated, rejected, or accepted. After a period of sometimes maddening activity, the dissections cease and employees sit back content, filled with energy and commitment. But it is the participation process that generates the reality to which they then make their commitment. As physicist Fred Wolf said: "According to the quantum rules, we cannot ever know and experience simultaneously all that is in principle knowable. . . . One thing is clear, though: self plays a role in what is seen to be not-self" (1989, 80-81).

Participation, ownership, subjective data—each of these organizational insights that I gain from quantum physics, whether I take them literally or metaphorically, quickly return me to a central truth. A quantum universe is enacted only in an environment rich in relationships. Nothing happens in the quantum world without something encountering something else. Nothing *is* independent of the relationships that occur. I am constantly creating the world—evoking it, not discovering it—as I participate in all its many interactions. This is a world of process, not a world of things.

Physicists have had a head start in becoming oriented to this new world of process. They pay attention to events and interactions rather than to things, thus becoming—in Gary Zukav's extended metaphor of the Wu Li Masters—observers of the dance (1979). But for us—as we sit in our offices, structured into rigid relationships, besieged with stacks of data that accumulate daily, armed with our complex formulae of interpretation—we have a long way to go before we can move onto that dance floor. It seems a distant vision, this belief in nonobjectivity, this realization that information is partly our creation, as is all of reality.

It makes me wonder how we will design our organizations in the future. As we struggle with the designs that will replace bureaucracy, we must invent organizations where process is allowed its varied-tempo dance, where structures come and go as they support the process that needs to occur, and where form arises to support the necessary relationships.

Physicists struggle with a similar dilemma when they try to diagram reactions between "things" that are not things until they are engaged with one another. There have been different ways of drawing the reactions by which particles appear, change, and participate in the creation of other particles. In two examples, Feynman diagrams and S-matrices, lines converge from different points, forming new lines that go off in other directions. The elaborate lattices of these drawings reinforce the idea that particles are best understood, not as objects, but as occurrences, as temporary states in a network of reactions that go on and on (see Bubble Chamber photograph of particle interactions on page 79).

Without understanding the science in detail, I have been intrigued by some of the concepts in S-matrix diagrams (the *S* stands for scattering). These diagrams represent a way of modeling the dynamic lives of high-energy particles (hadrons) and how they are able to manifest into several different forms, depending on the energy available. I have spent hours staring at these diagrams, knowing they have something to teach me about organizational structure and how we might chart roles and relationships differently (see Capra 1976, 1982; and Zukav 1979).

The first thing that intrigues me about these diagrams is the concept of

---

S-matrix diagrams are symbolic representations of subatomic particle interactions that occur within a given area. They are graphic depictions of particles as intermediate states in a network of interactions, where the energy present in any particle can combine with other energy sources to create new particles. The general area of interaction is depicted by the circle. The individual lines are not particles, but "reaction channels" through which energy flows.

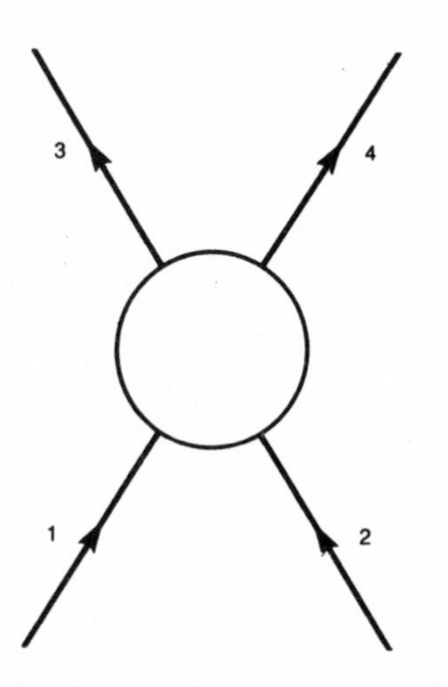

Particles 1 and 2 go into the collision area (the circle), and particles 3 and 4 come out. There is no top or bottom to these diagrams. With each rotation, new combinations of energy are possible that will emerge as different particles.

"reaction channels." In the diagrams, lines of direction converge into a collision circle, from which other lines emerge. Each of these lines has a particle name attached to it, but the lines are best understood not as particles, not as things, but as "reaction channels," where energy meets and takes temporary form. Several different particles can emerge in the reaction channels, depending on the amount of energy that is generated in the interactions.

In a traditional organizational chart, where we draw lines to connect roles, it would be a breakthrough to think of the lines as reaction channels, lines along which energy was transferred to facilitate the creation of new things. But S-matrices stretch my thinking in bigger ways than this because they demand that I

A "neutron" here is a reaction channel that can be formed by the combined energies of a proton and a negative pion.

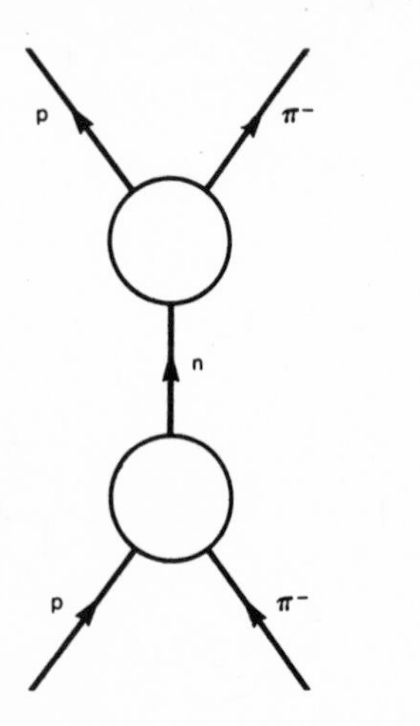

Given more energy, the same neutron channel could be formed by other particle combinations.

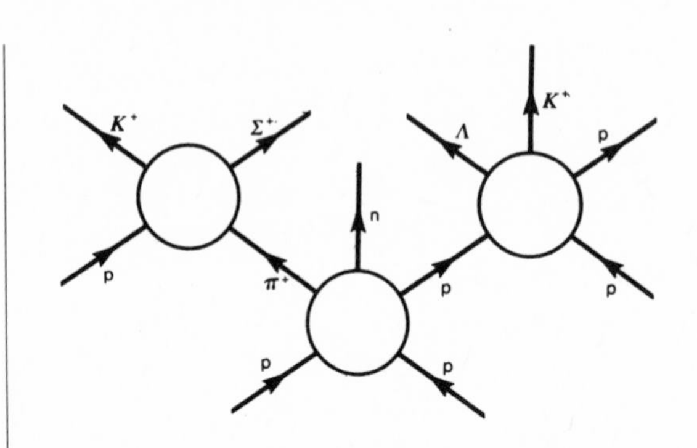

This S-matrix network of interactions illustrates the on-going exchange of energy which temporarily manifests as different particles. Particles exist as patterns of relationships; they can only be defined in terms of one another. It is impossible to describe any one particle as first or last; what is important is the direction and amount of energy available in any interaction.

stop thinking of roles or people as fixed entities.

A hadron is defined by its energy and by the network of relationships through which it exchanges energy. This is true of all subatomic particles that, in Capra's words, "are not separate entities but interrelated energy patterns in an ongoing dynamic process. These patterns do not 'contain' one another but rather 'involve' one another . . ." (1982, 94). These particles are described as a *tendency* to participate in various reactions, a definition which honors the dynamic qualities of their existence. With S-matrix diagrams, physicists are able to describe (mathematically) the processes of continual transformation, of emergence, decay, and the new formations that characterize high-energy particles. The result is an intriguing network of interactions, a structure of processes and potential relationships.

If I apply this to the roles and relationships described in traditional organization charts, I come up with some different ways of thinking about people. I cannot describe a person's role, or his or her potential contribution, without understanding the network of relationships and the energy that is required to create the work transformations that I am asking from that person. No longer can I define the person only in terms of his or her authority relationship to me. I need to be able to conceptualize the pattern of energy flows that are required for that person to do the job. If I can do this, I then see the person as a conduit for organizational energy, as the place where sufficient resources meet to make something happen. It gives me a very different perspective on what I must do to support that person and what is required to make the whole organization work at transformative energy levels.

S-matrix diagrams can also be rotated, thereby altering the reactions among the particle players. No one particle is the basic element or causative agent. Each has the capacity to interact with another and produce different outcomes. Rotating the diagrams changes the roles played by the different energies; what was a force influencing a reaction can, by turning the diagram, become a reaction channel influenced by other forces. Hierarchy and permanence are not what is important; rather, the presence of reaction channels for the exchange of energy is what matters.

Wouldn't it be interesting to draw an organization borrowing from S-matrix images? Is it possible to think about roles in this way, as minimally bounded, as focused on interactions and energy exchanges? Any role could be understood both as a reaction channel in which forms (specific tasks or accountabilities) appeared and as a generative force, capable of contributing its energy to others. In such an organization, we would draw structures to emphasize the interactions we needed, rather than isolated individuals, and we would want more information on the organization's capacity to facilitate energy flows. Our attention would be directed to the energy required to achieve a desired outcome.

To do this, to diagram the transformative energy of an organization or project, we would need to sharpen our understanding of the elements that create organizational energy—staff, time, resources, education, information, etc.—and be wiser about assessing their relative impact on the outcomes we desired. But if we succeeded in thinking about organizations in this way, we could begin to create organizations of process and relationships, quantum organizations that worked more effortlessly in this strange universe.

Heisenberg describes the world of modern physics as one divided not "into different groups of objects but into different groups of connections." What is distinguishable and important, he says, are the kinds of connections. This is the world in which we design and manage organizations, and there undoubtedly are important images from physics to challenge our images of organizations.

Perhaps these are just the ramblings of one whose mind has gone fuzzy (like all quantum phenomena) from trying to understand quantum physics. But there *is* an urgent challenge to create organizations that respond to this new world of relationships in which we act as grand evocateurs of reality. Our old views constrain us. They deprive us from engaging fully with this universe of potentialities.

When I think of all those wave functions filling space, rich in potentials, accumulating more and more possibilities as they fan out, I wonder why we limit ourselves so quickly to one idea or one structure or one perception, or to the idea that "truth" exists in objective form. Why would we stay locked in our belief that there is one right way to do something, or one correct interpretation to a situation, when the universe welcomes diversity and seems to thrive on a multiplicity of meanings? Why would we avoid participation and worry only about its risks, when we need more and more eyes to evoke reality? Why would we resist the rich visions and strong futures that emerge when we come together to create the world? Why would we ever choose rigidity or predictability when we have been invited to be part of the generative processes of the cosmos? And why would we ever peer into that box expecting a dead cat, when just by our powers of observation we could bring that cat to life?

"To live in an evolutionary spirit means
to engage with full ambition and without
any reserve in the structure of the present,
and yet to let go and flow into a new
structure when the right time has come."

—Erich Jantsch

CHAPTER 5

Change, Stability, and Renewal: The Paradoxes of Self-Organizing Systems

One day when a child, I stood beneath a swing frame that towered above me. Another child, older than me, told me of the time a girl had swung and swung until, finally, she looped over the top. I listened in silent awe. She had done what we only dreamed of doing, swung so uncontrollably high that finally not even gravity could hold her.

I think of this apocryphal story as I sit now in a small playground, watching my youngest son run from one activity to another. He has climbed, swung and jumped, whirled around on a spinning platform, and wobbled along a rolling log until, laughing, he loses his balance. Now he is perched on a teeter-totter, waiting to be bumped high in the air when his partner crashes to the ground. Everywhere I look, there are bodies in motion, energies in search of adventure.

It seems that the very experiences these children seek out are ones we avoid: disequilibrium, novelty, loss of control, surprise. These make for a good playground, but for a dangerous life. We avoid these things so much that if an organization were to take the form of a teeter-totter, we'd brace it up at both ends, turning it into a straight plank. But why has equilibrium become such a prized part of adult life? Why are we afraid of what happens if our boat gets rocked? Is it that we prefer balance to change? Does equilibrium feel more secure?

Sometimes, to clear up a confusing concept, it helps me to return to the accepted definition of the word. So I open the *American Heritage Dictionary* to learn about equilibrium: "Equilibrium. 1. A condition in which all acting influences are canceled by others resulting in a stable, balanced, or unchanging system. 2. Physics. The condition of a system in which the resultant of all acting forces is zero. . . . 4. Mental or emotional balance; poise."

I am surprised by the negativity of the first two definitions. A condition in which the result of all activity is zero? Why, then, do we desire equilibrium so much, or use the same word to describe mental and emotional well-being? In my own life, I don't experience equilibrium as an always desirable state. And I don't believe it is a desirable state for an organization. Quite the contrary. I've observed the search for organizational equilibrium as a sure path to institutional death, a road to zero trafficked by fearful people. Having noticed the negative effects of equilibrium so often, I've been puzzled why it has earned such high status. I now believe that it has to do with our outmoded views of thermodynamics.

Equilibrium is a result of the workings of the Second Law of Thermodynamics. Though we may not know what this law states, we act on its assumptions daily. My son learned it in fourth-grade physics as the "laziness law"—the tendency of closed systems to wear down, to give off energy that can never be retrieved. Ecologist Garrett Hardin aptly paraphrases this law: "We're sure to lose" (in Lovelock 1987, 124). Life goes on, but it's all downhill.

In classical thermodynamics, equilibrium is the end state in the evolution of isolated systems, the point at which the system has exhausted all of its capacity for change, done its work, and dissipated its productive capacity into useless entropy. (Entropy is an inverse measure of a system's capacity for change. The more entropy there is, the less the system is capable of changing.) At equilibrium, there is nothing left for the system to do; it can produce nothing more. If the universe is a closed system (there being nothing outside the universe to influence it), then it too must eventually wind down and reach equilibrium. It will become a place where, in the words of scientists Peter Coveney and Roger Highfield, "entropy and randomness are at their greatest, in which all life has died out" (1990, 153).

The Second Law of Thermodynamics only applies to isolated and closed systems—to machines, for example. The most obvious exception to this law is *life*, open systems that engage with their environment and continue to grow and evolve. Yet both our science and culture have been profoundly affected by the images of degeneration contained in classical thermodynamics. When we see decay as inevitable, or society as going to ruin, or time as the road to inexorable death, we are unintentional celebrants of the Second Law. James Lovelock, biologist and author of the Gaia hypothesis, says the laws of thermodynamics "read like the notice at the gates of Dante's Hell" (1987, 123).

In a universe that is on a relentless road to death, we live in great fear. Perhaps we become so fearful of change because it uses up valuable energy and leaves us only with entropy. Staying put or keeping in balance are our means of defense against the eroding forces of nature. We want nothing to rock the boat because only decline awaits us. Any form of stasis is preferable to the known future of deterioration.

But in venerating equilibrium, we hide from the processes that foster life. It is both sad and ironic that we have treated organizations like machines, acting as though they were dead when all this time they've been living, open systems capable of self-renewal. We have magnified the tragedy by treating one another as machines, believing the only way we could motivate others was by pushing and prodding them into action, overcoming their entropy by the sheer force of our own energy. But here we are, living beings in living systems in a universe that continues to grow and evolve. Can we dump these thermodynamics and get to the heart of things? Can we respond to *life* in organizations and discard the deathwatch? Can we cooperate with living systems and stop our clumsy attempts to restrain change or to suppress disturbances?

Equilibrium is neither the goal nor the fate of living systems, simply because as open systems they are partners with their environment. The study of these systems, begun with Prigogine's prize-winning work (1980), has shown that open systems have the possibility of continuously importing free energy from the environment and of exporting entropy. They don't sit quietly by as their energy dissipates. They don't seek equilibrium. Quite the opposite. To stay viable, open systems maintain a state of non-equilibrium, keeping the system off balance so that it can change and grow. They participate in an active exchange with their world, using what is there for their own renewal. Every organism in nature, including us, behaves in this way.

In the past, systems analysts and scientists studied open systems primarily focusing on the overall *structure* of the system. This route led away from observing or understanding the processes of change and growth that make a system viable over time. Instead, analysts went looking for those influences that would support *stability*, which is the desired trait of structures. Feedback loops were monitored as a way of maintaining system stability. Regulatory or negative feedback loops served this function well, signalling departures from the norm. As managers watched for sub-standard performance, they could make corrections and preserve the system at its current levels of activity.

But there is a second type of feedback loop—positive ones that amplify

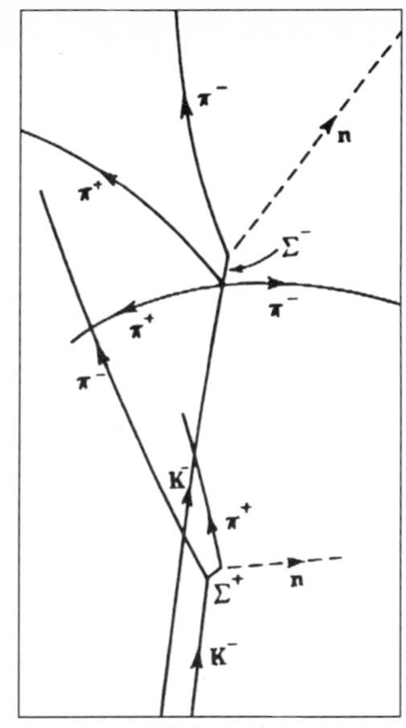

Lawrence Berkeley Laboratory,
University of California

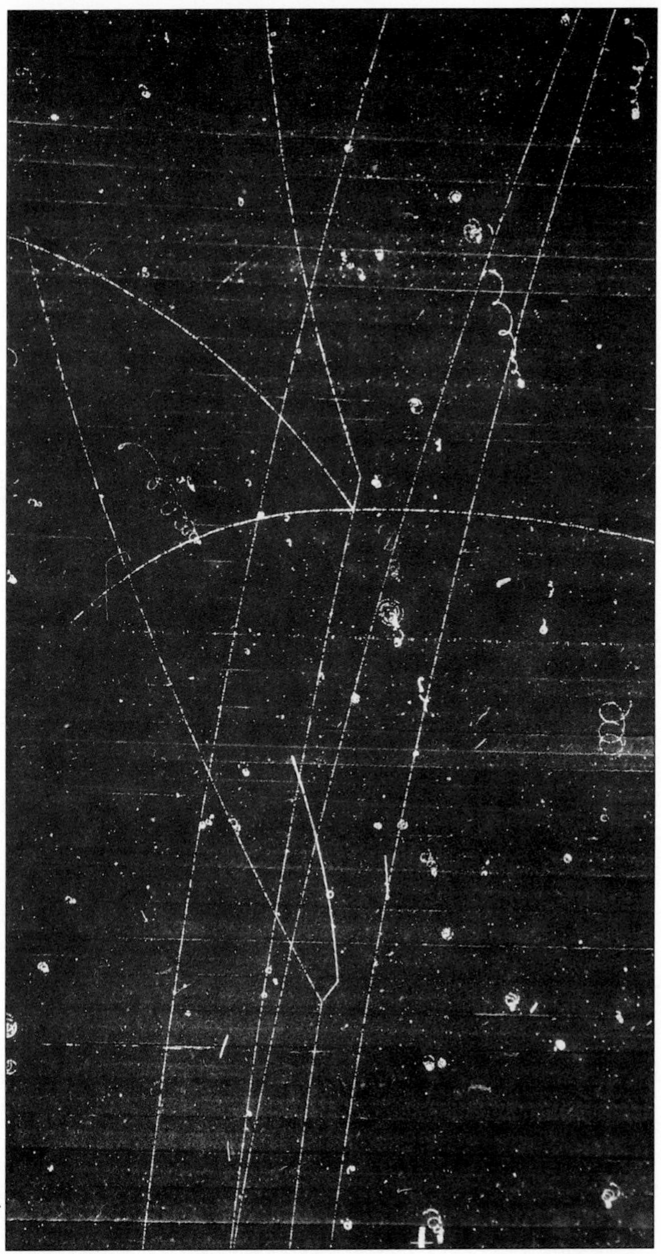

A beam of K mesons (seen as straight lines) enters the hydrogen atmosphere of the bubble chamber at the bottom of the photo. Two of the K mesons meet up with protons (not visible), and in less than one/billionth of a second, the energy exchange results in four interactions, producing ten visible particles. The negative K meson (K-) at bottom right transforms into a sigma plus (Σ+) particle. This sigma particle in turn transforms into a neutron (n), which decays off to the right of the picture, a positive pion (π+), and a negative pion (π-) both of which then decay.

The second K- meson develops a reverse charge, and in collision with a proton, transforms into a positive Pi (π+) particle, two pions and a negative Sigma particle. This Sigma then decays into a neutron and a minus pi (π-).

The spirals and curlicues seen scattered throughout the photo are decaying particles, most likely electrons; as they lose energy, they are pulled into the magnetic field of the chamber.

Fractal shapes, created by repeated interations of non-linear equations, appear everywhere. Their beauty and variety emerge as a result of two contradictory processes: total freedom for the equations to evolve as they will, with no moment-to-moment prediction possible; yet a pre-determined final shape described by the initial parameters.

A fractal object repeats a similar pattern or design at ever-smaller levels of scale. No matter where you look, the same pattern will be evident. In any fractal object, we are viewing a simple organizing structure that creates unending complexity.

These Julia Set fractal images represent the chaotic yet orderly ramblings of a

---

"It was always here. . .
waiting to be seen. . . .
Why do the shapes occur
where they do?. . .
They dust the complex plane
like stars and galaxies clustering in
ever-higher aggolomerations,
in an infinitude of shapes
and levels."

—Dan Kalikow

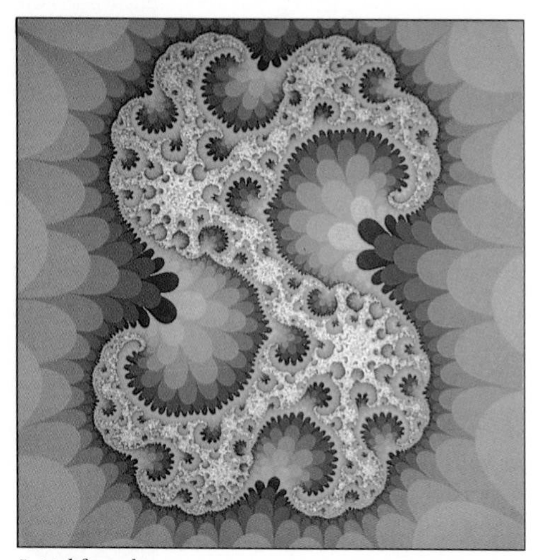

Initial fractal

magnification: 265

magnification: 1 million

simple, non-linear formula, ($z_{n+1}=z_{n^2+c}$) *as it developed over millions of iterations. Color values are set by the computer programmer to correspond to the different numeric values created by the equation's calculations.*

In this series of six photos, we look deeper and deeper into the structure of the initial fractal object. Even at magnifications of one trillion, self-similar patterns are evident. We stopped peering in at this magnification, but we could have explored even deeper, never running out of regions to explore, always seeing these self-similar shapes.

Fractals provide a glimpse of infinity that is well-bounded; of simplicity feeding back on itself to create beautiful complexity.

Photo by Lifesmith Classic Fractals

magnification: 1 billion

magnification: 40 billion

magnification: 1 trillion

"Within its deep infinity
I saw ingathered,
and bound by love
in one volume,
the scattered leaves
of all the universe."

—Dante

Clouds are developed by the repetition of similar shapes, repeated at many different levels of scale. It is fascinating to explore the fractal nature of clouds from an airplane window.

Photo by Royce Bair

Many vegetables are fractal. The dominant shape of broccoli can be seen even in the individual elements that make up a floret.

Photo by
Steve Tregeagle

Because of the fractal nature of ferns, it is possible to generate them on computers, using a few numeric values that describe the basic dimensions of the fern (affine transformations). This is the chaos game, where simple rules combine with chaos to form predictable order.

Photo by Lifesmith Classic Fractals

In this scene of the Grand Canyon, other smaller canyons are evident in the foreground. Landscape photographers, intrigued by repetitive patterns, often capture fractal qualities in their photographs. Photo by Frank Jensen

Spiral patterns, so common in nature, emerge from the dance of order and chaos.

The Belousov-Zhabotinsky Reaction: a Self-Organizing Chemical Solution. Observed in a dish, chemicals are subjected to changing conditions, beginning upper left. They react to the new information, amplifying and changing it through a basic interactive process. The chemical solution moves from chaos, into new order, ending lower right. Spiral structures emerge as a result of the solution's self-organizing into new forms.

Photo by A. Winfree

Hurricane Elena, from the space shuttle Discovery, *September 1, 1985. The scroll formation is created as massive energy self-organizes into a formidable weather system. Many galaxies also exhibit this scroll form. The same iterative model that produces the Belousov-Zhabotinsky reaction is at work in these enormous energy systems.*

Photo by National Weather Service

Temple Megaliths of Tarxien, Malta, c. 3000 B.C. Spirals appear in art all over the world, beginning with Paleolithic cave drawings. Psychologist Carl Jung believes the spiral is an archetype embedded deep within the collective unconscious. The spirals on this Maltese temple sprout leaves to depict the life force inherent in the on-going dance of chaos and order.

Photo by Marija Gimbutas

Three-winged Bird:A Chaotic Strange Attractor, from the work of Mario Markus and Benno Hess.

Max-Planck Institute, Dortmund, Germany

responses and phenomena. These loops use information differently, not to regulate, but to amplify into troublesome messages, like the ear-piercing shrieks of microphones caught in a positive feedback loop. In these loops, information increases and disturbances grow. The system, unable to deal with so much magnifying information, is being asked to change. For those interested in system stability, amplification is very threatening, and there is a need to quell it before eardrums burst.

For many years scientists failed to notice the role positive feedback and disequilibrium played in moving a system forward. In trying to preserve things as they were, in seeking system stability, they failed to note the internal processes by which open systems accomplish growth and adaptation.

It was not until the element of time was introduced in Prigogine's study of thermodynamics that interest turned from system structures to system dynamics. His work, and those who built on it, dramatically expanded our awareness of how open systems use disequilibrium to avoid deterioration. Looking at the dynamics of open systems over time, scientists were able to see the effects of energy transformations that had not previously been observed. Entropy, that fearful measure of a system's demise, was still being produced, sometimes in great quantities. But instead of simply measuring *how much* entropy was present, scientists could also note *what happened* to it—how quickly it was produced and whether it was exchanged with the environment.

Once it was noted that systems were capable of exchanging energy, taking in free energy to replace the entropy that had been produced, scientists realized that deterioration was not inevitable. Disturbances could create disequilibrium, but disequilibrium could lead to growth. If the system had the capacity to react, then change was not necessarily a fearsome opponent. To understand the world from this perspective, scientists had to give up their views on decay and dissipation. They had to transform their ideas about the role of disequilibrium. They had to develop a new relationship with disorder.

Prigogine's work on the evolution of dynamic systems demonstrated that disequilibrium is the necessary condition for a system's growth. He called these systems *dissipative structures* because they dissipate their energy in order to recreate themselves into new forms of organization. Faced with amplifying levels of disturbance, these systems possess innate properties to reconfigure themselves so that they can deal with the new information. For this reason, they are frequently called self-organizing or self-renewing systems. One of their distinguishing features is system *resiliency* rather than stability.

There are many startling examples of dissipative structures in chemical reactions. One of the most often described is the behavior exhibited by chemical clocks. A chemical clock is a mixture that oscillates between two different states. In the normal scheme of things, we expect that when chemicals are mixed together, they form a substance where the molecules of the chemicals are evenly distributed. If one chemical is blue and the other is red, we expect that the mixture will be purple. This is, in fact, the case when the chemical clock is at equilibrium and no reactions are taking place. But when change is introduced into this dissipative structure by adding more of the two chemicals into the dish (or adding heat or different chemicals), disequilibrium occurs, and the system behaves in a manner that defies normal expectations. Instead of more purple, the substance begins to pulsate, first red, then blue, with a predictable cycle that has earned these solutions the name of "clocks." To keep the color pulsations, the mixture must continue to be infused with more disturbances in the form of chemicals or conditions. If these decrease below a certain threshold, equilibrium returns and the solution returns to its purple state.

These chemical reactions use a great deal of energy. Entropy has increased during this reaction, but it has been exchanged with the environment for usable energy. As long as the system stays open to the environment, and matter and energy continue to be exchanged, the system will avoid equilibrium and remain, instead, in these "evanescent structures" that exhibit "exquisitely ordered behavior" (Coveney and Highfield 1990, 164).

There are many examples of self-organizing chemical clocks that exhibit extraordinary behavior. One of the most beautiful is the Belousov-Zhabotinsky reaction, where the chemicals, in response to changes in temperature and mix, form into swirling spiral patterns that rival the beauty of a Ukranian Easter egg. Here, the system responds to novelty and change by creating an entirely new level of exquisite organization.

The scrolls that emerge in the Belousov-Zhabotinsky reaction are similar to the scroll formations that appear in many other places, both in nature and in art. "The spiral is one of nature's basic forms of design," writes photographer Andreas Feininger (1986, 124). Some scientists have wondered if spiral forms in art describe a common, deep experience of change, change that leads to dissipation and then to a new ordering. We see such spiral patterns on weather maps that track hurricanes. We live in a spiral-shaped galaxy; in fact, astronomers studying our type of disk galaxy have concluded that the same iterative model used in the simple Belousov-Zhabotinsky chemical reaction applies to the scroll formation of ancient star clusters. John Briggs, a science writer, and his writing colleague, physicist David Peat, note the scroll images found so frequently in art, particularly noting the interlocking scroll patterns found in early motifs throughout the world: "Could such a collective wisdom perhaps be expressing its intuitions of the wholeness within nature, the order and simplicity, chance and predictability that lie in the interlocking and unfolding of things?" (1989, 142-43). (See illustrations, page 85.)

The self-organizing dynamics exhibited by chemical clocks are evident in all open systems. These dynamics apply to such a broad spectrum of phenomena that they unify science across the domains of many disciplines. But, more importantly, they give us a new picture of the world, they "let us feel the *quality* of a world which gives birth to ever new variety and ever new manifestations of order against a background of constant change" (Jantsch 1980, 57).

One thing that I find especially intriguing about self-renewing systems is their relationship with their environment. It feels new to me. In organizations, we typically struggle against the environment, seeing it as the source of disruption and change. We tend to insulate ourselves from it as long as possible in an effort to preserve the precious stability we have acquired. Even though we know we must become responsive to forces and demands beyond the boundaries of our organizations, we still focus our efforts on maintaining the strongest defensive structure possible. We experience an inherent tension between stability and openness, a constant tug-of-war, an either/or. But as I read about self-renewing structures, these dualities feel different. Here are structures that seem capable of maintaining an identity while changing form. So how do they do it?

Part of their viability comes from their internal capacity to create structures that fit the moment. Neither form nor function alone dictates how the system is constructed. Instead, form and function engage in a fluid process where the system may maintain itself in its present form or evolve to a new order. The system possesses the capacity for spontaneously emerging structures, depending on what is required. It is not locked into any one form but instead is capable of organizing information in the structure that best suits the present need.

We are beginning to see organizations that tap into this property of self-organizing or self-renewing systems. Some theorists have termed these "adaptive organizations," where the task determines the organizational form (Dumaine 1991). In a separate but related example are those corporations structured around core competencies, as described by C. K. Prahalad and Gary Hamel (1990). Both types of organizations avoid rigid or permanent structures and instead develop a capacity to respond with great flexibility to external and internal change. Expertise, tasks, teams, and projects emerge in response to a need. When the need changes, so does the organizational structure.

But an organization can only exist in such a fluid fashion if it has access to new information, both about external factors and internal resources. It must constantly process this data with high levels of self-awareness, plentiful sensing devices, and a strong capacity for reflection. Combing through this constantly changing information, the organization can determine what choices are available, and what resources to rally in response. This is very different from the more traditional organizational response to information, where priority is given to maintaining existing operating forms and information is made to fit the structure so that little change is required.

While a self-organizing system's openness to new forms and new environments might seem to make it too fluid, spineless, and hard to define, this is not the case. Though flexible, a self-organizing structure is no mere passive reactor to external fluctuations. As it matures and stabilizes, it becomes more efficient in the use of its resources and better able to exist within its environment. It establishes a basic structure that supports the development of the system. This structure then facilitates an insulation from the environment that protects the system from constant, reactive changes.

For example, in the early stages of an ecosystem, those species predominate that produce large numbers of offspring, most of which die. These species are vulnerable to changes in the environment, and there is an inefficient use of energy in the production and death of so many offspring. At this early stage, the environment exerts extreme pressure, playing a dominant role in the selection of species. But as the ecosystem matures, it develops an internal stability, a resiliency to the environment that, in turn, creates conditions that support more efficient use of energy and protection from environmental demands. The system develops enough stability to support mammals, which produce far fewer offspring than lower species, but which can now survive because of the system's well-developed structure (see Jantsch 1980, 140ff; Margalef 1975).

What occurs in these systems is contrary to our normal way of thinking. Openness to environmental information over time spawns a firmer sense of identity, one that is less permeable to externally induced change. Some fluctuations will always break through, but what comes to dominate the system over time is not environmental influences, but the self-organizing dynamics of the system itself. High levels of autonomy and identity result from staying open to information from the outside.

I say this is contrary thinking because we often practice a reverse belief—that to maintain our identity, our individuality, we must protect ourselves from the demands of external forces. We tend to think that isolation and clear boundaries are the best way to maintain individuality. But in the world of self-organizing structures, we learn that useful boundaries develop through openness to the environment. As the process of exchange continues between system and environment, the system, paradoxically, develops greater freedom from the demands of its environment.

Companies organized around core competencies provide a good example of how an organization can obtain internal stability that leads both to well-defined boundaries and to openness over time. A business that focuses on its core competencies identifies itself as a portfolio of skills rather than as a portfolio of business units. It can respond quickly to new opportunities because it is not locked into the rigid boundaries of preestablished end products or businesses. Such an organization is both sensitive to its environment, and resilient from it. In deciding on products and markets, it is guided internally by its competencies, not just the attractiveness or difficulty of a particular market. The presence of a strong competency identity makes the company less vulnerable to environmental fluctuations; it develops an autonomy that makes it unnecessary to be always reactive.

Yet such companies are remarkably sensitive to their environment, staying wide open to new opportunities and ventures that welcome their particular skills. They also develop capacities to shape the environment, creating markets where none existed before. In the assessment of Prahalad and Hamel, companies focused on core competencies are able to "invent new markets, quickly enter emerging markets, and dramatically shift patterns of customer choice in established markets" (1990, 80).

These companies highlight a principle that is fundamental to all self-organizing systems, that of *self-reference*. In response to environmental disturbances that signal the need for change, the system changes in a way that remains consistent with itself in that environment. The system is autopoietic, focusing its activities on what is required to maintain its own integrity and self-renewal. As it changes, it does so by referring to itself; whatever future form it takes will be consistent with its already established identity. Changes do not occur randomly, in any direction. They always are consistent with what has gone on before, with the history and identity of the system. This consistency is so strong that if a biological system is forced to retreat in its evolution, it does so along the same pathway. The system, in Jantsch's terms, "keeps the memory of its evolutionary path" (1980, 1; see also 49).

Self-reference is what facilitates orderly change in turbulent environments. In human organizations, a clear sense of identity—of the values, traditions, aspirations, competencies, and culture that guide the operation—is the real source of independence from the environment. When the environment demands a new response, there is a reference point for change. This prevents the vacillations and the random search for new customers and new ventures that have destroyed so many businesses over the past several years.

Another characteristic of self-organizing systems is their stability over time. They are often referred to as globally stable structures. Yet when we speak of the stability of mature self-organizing systems, we are referring only to a quality of the whole system. In fact, this global stability is maintained by another paradoxical situation, the presence of many fluctuations and instabilities occurring at local levels throughout the system. To use the example of an ecosystem again, any mature ecosystem experiences many changes and fluctuations at the level of individuals and species. But the total system remains stable, capable of developing its own rhythm of growth and lessening the impact on the system of such outside disturbances as climatic change (Jantsch 1980, 142). Small, local disturbances are not suppressed; there is no central command control that prohibits small, constant changes. The system allows for many levels of autonomy within itself, and for small fluctuations and changes. By tolerating these, it is able to preserve its global stability and integrity in the environment.

Jantsch notes the profound teaching embedded in these system characteristics: "The natural dynamics of simple dissipative structures teach the optimistic principle of which we tend to despair in the human world: *the more freedom in self-organization, the more order*" (1980, 40; italics added).

Here is another critical paradox: The two forces that we have always placed in opposition to one another—freedom and order—turn out to be partners in generating viable, well-ordered, autonomous systems. If we allow autonomy at the local level, letting individuals or units be directed in their decisions by guideposts for organizational self-reference, we can achieve coherence and continuity. Self-organization succeeds when the system supports the independent activity of its members by giving them, quite literally, a strong frame of reference. When it does this, the global system achieves even greater levels of autonomy and integrity.

In addition to these tantalizing paradoxes, there seems to be yet another important teaching for organizations in the behavior of self-organizing systems. Under certain conditions, when the system is far from equilibrium, creative individuals can have enormous impact. It is not the law of large numbers, of favorable averages, that creates change, but the presence of a lone fluctuation that gets amplified by the system. Through the process of autocatalysis, where a small disturbance is fed back on itself, changing and growing, exponential effects can result. "The ability of a system to amplify a small change is a creative lever," note Briggs and Peat (1989, 145).

It is natural for any system, whether it be human or chemical, to attempt to quell a disturbance when it first appears. But if the disturbance survives those first attempts at suppression and remains lodged within the system, an iterative process begins. The disturbance increases as different parts of the system get hold of it. Finally, it becomes so amplified that it cannot be ignored. This dynamic supports some current ideas that organizational change, even in large systems, can be created by a small group of committed individuals or champions.

Certain conditions support this process of change in both molecules and people. The revolutionaries cannot be isolated from one another. They must keep a firm grasp on their intentions and not let them be diffused into the larger system too early. And they must have links to other parts of the system. While this prescription reads like either a 1960s handbook on revolution or more recent texts on organizational change, it is, in fact, the governing principle by which self-organizing structures evolve. In some ways, it is humbling to realize that we have not invented our strategies for change; we have merely discovered them.

Self-organizing systems driven to change by this amplification process are given opportunities for creative reordering. If amplifications have increased to the level where the system is at maximum instability (a crossroads between death and transformation known technically as a bifurcation point), the system encounters a future that is wide open. No one can predict which evolutionary path it will take. Evolution itself is not constraining; the system is free to seek out its own optimal solution to the current environment. Except for honoring the principle of self-reference, the system has no predetermined course. At the bifurcation point, "such systems seem to 'hesitate' among various possible directions of evolution," Prigogine and Stengers state; "a small fluctuation may start an entirely new evolution that will drastically change the whole behavior of the macroscopic system" (1984, 14).

I can think of several organizations, particularly market-oriented ones, that brag about how a customer inquiry or the suggestion of an employee directed them into new product lines that became very successful. There was no pre-planning, no long-range strategic objectives, that led them into these markets. Just the creativity of one or a few individuals who succeeded in getting the attention of the organization and then watched the process amplify itself into a new, unexpected direction for the company.

The openness and creativity that influence a system's evolution will also affect the evolution of the environment. Self-organizing systems do not simply take in information; they change their environment as well. No part of the larger system is left unaffected by changes that occur someplace within it. Known to scientists as co-evolution, organizational theorist William Starbuck wrote years ago about a similar process in organizations: "The constraints imposed by environmental properties are not, in general, sufficiently restrictive to determine uniquely the characteristics of their organizational residents. . . . Organizations and their environments are evolving simultaneously toward better fitness for each other" (1976, 1105-06).

In this view of evolution, the system changes, the environment changes, and, some scientists argue, even the rules of evolution change. "Evolution is the result of self-transcendence at all levels. . . . [It] is basically open. It determines its own dynamics and direction. . . . By way of this dynamic interconnectedness, evolution also determines its own *meaning*" (Jantsch 1980, 14).

In the world of self-organizing structures, everything is open and susceptible to change. But change is not random or incoherent. Instead, we get a glimpse of systems that evolve to greater independence and resiliency because they are free to adapt, and because they maintain a coherent identity throughout their history. Stasis, balance, equilibrium—these are temporary states. What endures is process—dynamic, adaptive, creative.

If an open system seeks to establish equilibrium and stability through constraints on creativity and local changes, it creates the conditions that threaten its survival. We see evidence of this in many organizations, but among the most dramatic have been the many ecological messes we've created because small, natural fluctuations were dampened in our attempt to manage a wildlife area or population. In Yellowstone, for example, human-imposed stability thwarted the natural, small fluctuations of fires for many years. The result was a fragile equilibrium completely vulnerable to the cataclysm of fire that consumed the brush and dead trees that had accumulated because of man's desire for local control.

The more I read about self-renewing systems, the more I marvel at the images of freedom and possibility they evoke. This is a domain of independence and interdependence, of processes that support forces we've placed in opposition—change and stability, continuity and newness, autonomy and control—and all in an environment that tests and teases and disturbs and, ultimately, responds to changes it creates by changing itself. The traditional contradictions of order and freedom, change and stasis, being and becoming—these all whirl into a new image that is very ancient—the unifying dance of the great polarities of the universe.

The world of dissipative structures is rich in knowledge of how the world works, of how order is sustained by growth and change. This is very new territory for us, and it is difficult to avoid exploring it with our well-trained linear minds, looking for immediate applications and techniques that apply directly to the systems we live in. But if that is all we do (and I have done a bit of it myself in this chapter), we diminish the impact of this new landscape. If we plod across this new territory, heads down, our attention focused on specific features of the land, we may fail to look up and take in the whole of things here. We may fail to sense how life is maintained and how things work together, and we may fail to see the unifying process that embraces great paradoxes.

I find pleasure in letting these new concepts swirl about me. Like clouds, they appear, transform, and move on. Clouds themselves are self-organizing, changing into thunderstorms, hurricanes, or rain fronts with the influx of atmospheric energy or foreign particles. We are capable of similar transformations when we trust that new thoughts and ideas can self-organize in the environment of our minds and our organizations. And we would do well to take clouds more seriously. They are spectacular examples of strange and unpredictable systems, structured in ways we never imagined possible. "After all, how do you hold a hundred tons of water in the air with no visible means of support? You build a cloud" (Cole 1984, 38).

"An organization that creates information
is nothing but an organization that allows a
maximum of self-organizing order
or information out of chaos."

—Ikujira Nonaka

CHAPTER 6

The Creative Energy of the Universe—Information

Why is there such an epidemic of "poor communications" within organizations? In every one I've worked in, employees have ranked it right at the top of major issues. Indeed, its appearance on those lists in the past years became so predictable that I grew somewhat numb to it. Poor communication was a superficial diagnosis, I thought, that covered up other, more specific issues. Over the years, I developed a conditioned response to "communications problems" the minute they were brought up. I disregarded the category. I started pushing people to "get beyond" that catch-all phrase, to "give me more concrete examples" of communications failures. I believed I was en route to the "real" issues that would have nothing to do with communication.

Now I know I was wrong. My frustration with pat phrases didn't arise from people's lack of clarity about what was bothering them. They were right. They *were* suffering from information problems. Asking them to identify smaller, more specific problems was pushing them in exactly the wrong direction, because the real problems were big—bigger than anything I imagined. What we were all suffering from, then and now, is a fundamental misperception of information: what it is, how it works, and what we might expect from it.

The nub of the problem is that we've treated information as a "thing," as an inert entity to disseminate. Things are stable; they have dimensions and volume. You can get your hands around a thing. You can move it, track it, pass it back and forth. Things can be managed because they're so concrete. This "thing" view of information arose from several decades of information theory that treated information as a quantity, as "bits" to be transmitted and received. Information was a commodity to transfer from one place to another. The content, meaning, and purpose of information were ignored; they were not part of the theoretical construct (Gleick 1987, 255-56). Information theorists also focused on "noise"—those interferences that prevented smooth movement of the bits. Ideally, it was felt, information moved virginlike through the system, untouched by anything.

I believe it is information theory that has gotten us into trouble. We don't understand information at all.

What's curious about our misperceptions of information is that we all started out on a much higher plane of awareness. Remember playing "telephone" and being delighted and amazed at how the message got distorted with only a few players? At a young age, we knew information for its dynamic qualities, for its constantly changing aliveness. But when we entered organizational life, we left that perspective behind. We expected information to be controllable, stable, and useful for our purposes. We expected to be able to manage it.

In the universe new science is exploring, information is a very different "thing." It is not the limited, quantifiable, put-it-in-a-memo-and-send-it-out commodity with which we have become so frustrated. In new theories of evolution and order, information is a dynamic element, taking center stage. It is information that gives order, that prompts growth, that defines what is alive. It is both the underlying structure and the dynamic process that ensure life.

How can information be a dynamic or a structure rather than just content? A dramatic example of this, one that pushes our self-concept to the edge, is seen by asking: Who am I? Am I a body containing a mind, or a mind that has created a body? Am I a physical structure that processes information or non-physical information organizing itself into form?

Although we experience ourselves as a stable form, our body changes frequently. As physician Deepak Chopra likes to explain, our skin is new every month, our liver every six weeks; and even our brain, with all those valuable cells storing acquired knowledge, changes its content of carbon, nitrogen, and oxygen about every twelve months. Day after day, as we inhale and exhale, we give off what were our cells, and take in elements from other organisms to create new cells. "All of us," observes Chopra, "are much more like a river than anything frozen in time and space" (1990).

In spite of this exchange, we remain rather constant, due to the organizing function of the *information* contained in our DNA.

> At any point in the bodymind, two things come together—a bit of information and a bit of matter. Of the two, *the information has a longer life span than the solid matter it is matched with*. As the atoms of carbon, hydrogen, oxygen, and nitrogen swirl through our DNA, like birds of passage that alight only to migrate on, the bit of matter changes, yet there is always a structure waiting for the next atoms. In fact, DNA never budges so much as a thousandth of a millimeter in its precise structure, because the genomes—the bits of information in DNA—remember where everything goes, all 3 billion of them. This fact makes us realize that memory must be more permanent than matter. What is a cell, then? *It is a memory that has built some matter around itself, forming a specific pattern. Your body is just the place*

your memory calls home. (Chopra 1989, 87; italics added)

Jantsch describes the same phenomenon in dissipative structures, asking whether they should be understood as a material structure that organizes energy, or as an energy structure capable of organizing the flow of matter. "At higher levels of self-organization," he concludes, "a description will suggest itself which views energy systems manifesting themselves in the organization of material processes and structures" (1980, 35).

Information organizes matter into form, resulting in physical structures. The function of information is revealed in the word itself: in-*formation*. We haven't noticed information as structure because all around us are physical forms that we can see and touch and that beguile us into confusing the system's structure with its physical manifestation. Yet the real system, that which endures and evolves, is energy. Matter flows through it, assuming different forms as required. When the information changes (as when disturbances increase), a new structure materializes. Even a large structure like an ecosystem has been described similarly, as "an information system which manifests itself in the organization of matter," evolving as it accumulates information (Jantsch 1980, 141).

In a constantly evolving, dynamic universe, information is the fundamental ingredient, the key source of structuration—the process of creating structure. Something we cannot see, touch, or get our hands around is out there, organizing life. Information is managing us.

For a system to remain alive, for the universe to move onward, information must be continually generated. If there is nothing new, or if the information that exists merely confirms what is, then the result will be death. Isolated systems wind down and decay, victims of the laws of entropy. The fuel of life is new information—novelty—ordered into new structures. We need to have information coursing through our systems, disturbing the peace, imbuing everything it touches with new life. We need, therefore, to develop new approaches to information—not management but encouragement, not control but genesis. How do we create more of this wonderful life source?

Information is unique as a resource because of its capacity to generate itself. It's the solar energy of organization—inexhaustible, with new progeny emerging every time information meets up with itself. As long as there are senders and receivers linked together in a context, fertility abounds. All that is needed is freedom of circulation to guarantee new births. In fact, the greatest generator of information is chaos, where so much spawning of information goes on that researchers feel obliged to monitor every moment of the system's activity lest they miss something (Gleick 1987, 260).

Of course, this is exactly what we fear. We have no desire to let information roam about, to let it procreate promiscuously where it will, to create chaos. Our management task is to enforce control, to keep information contained, to pass it down in such a way that no procreation occurs. Information chastity belts are a central management function. The last thing we need is information running loose in our organizations. And there are good reasons for our stern, puritanical attitudes toward information: Misplaced information seems to have created enough horror stories to justify our frequent witch hunts.

But if information is to function as a self-generating source of organizational vitality, we must abandon our dark cloaks of control and trust in the principles of self-organization, even in our own organizations. Information is the source of order, an order we do not impose, but an order nonetheless. All of nature uses information this way. Can information, therefore, be used as an ordering mechanism for humanly-created organizations?

Information can serve such an organizational function only if organizations are living entities and respond to the same dynamics as open systems. A key question, then, is, are organizations alive, are they conscious, responsive entities? A new definition of consciousness—broader and more provocative—is emerging in some fields of science that can help frame an answer to these questions.

Prigogine was stimulated to think about consciousness when he observed a process of communication in certain chemical reactions. He concluded that even in "non-living" chemical solutions, communication occurs, generating order. In the chemical clocks he studied, the random mix of molecules became coordinated at a certain point. A murky gray solution, for example, suddenly would begin pulsing, first black, then white. In chemical clocks, all molecules act in total synchronization, changing their chemical identity simultaneously. "The amazing thing," Prigogine notes, "is that each molecule knows in some way what the other molecules will do at the same time, over relatively macroscopic distances. These experiments provide examples of the ways in which molecules communicate. . . . That is a property everybody always accepted in living systems, but in nonliving systems it was quite unexpected" (1983, 90).

If the capacity to deal with information, to communicate, defines a system as conscious, then the world is rich in consciousness, extending to include even those things we have classified as inanimate. Consciousness occurs in systems that do not even have an identifiable brain.

If we understand consciousness from the viewpoint of machine imagery, this makes no sense. If there is no identifiable part that handles thinking and communication, then there can be no such activity. In the past, we measured an organism's capacity for intelligence by counting the parts of its brain (or noting the lack of one). If crayfish have only 90,000 neurons, compared to our 10 billion or so, then certainly they aren't very smart. But then we discovered that crayfish are capable of doing far more than could have been predicted from our mechanistic models. One school of researchers working in artificial intelligence suggests that consciousness can't be discerned from the constituent parts of an entity. Instead, consciousness is a property that emerges when a certain level of organization is reached. Anything capable of self-organizing, therefore, possesses a level of consciousness. A well-ordered system is defined not by how many brain parts it has, but by how much information it can process. The greater the ability to process information, the greater the level of consciousness.

With this definition, organizations qualify as conscious entities. They also meet Gregory Bateson's (1980) criteria for "Mind." They have capacities for generating and absorbing information, for feedback, for self-regulation. In fact, information is an organization's primary source of nourishment; it is so vital to survival that its absence creates a strong vacuum. If information is not available, people make it up. Rumors proliferate, things get out of hand—all because people lack the real thing. Given the need for constant nourishing information, it is no wonder that "poor communication" inevitably appears so high on the problems list. Employees know it is the critical vital sign of organizational health.

We have lived for so long in the tight confines of bureaucracies—what Max De Pree, former CEO of Herman Miller, describes as "the most superficial and fatuous of all relationships"—that we need to learn how to live in a conscious organization, how to facilitate its intelligence. This requires an entirely new relationship with information, one in which we embrace its living properties. Not so that we open ourselves to indiscriminate chaos, but so that we facilitate aliveness and responsiveness. If we are seeking *resilient* organizations, a property prized in self-organizing systems, information needs to be our key ally.

Think about how we generally have treated information in the past. We've known it was important, but we've handled it in ways that have destroyed many of its life-giving properties. For one thing, we've taken disturbances and fluctuations and averaged them together to give us comfortable statistics. Our training has been to look for big numbers, important trends, major variances. Yet it is the slight variations—soft-spoken, even whispered at first—that we need to encourage. (Some of the recent work with statistics used in quality programs does emphasize the detection of these slight variations.)

Or we've taken conflicting information, rich with the possibility of moving us to new levels of understanding, and, instead, felt the need to play Solomon, to decide which piece of information or which position was correct. "Let's get to the bottom of this," we say, pointing our efforts dead into the ground—away from the conflicts that can move us toward the light, toward new, more complex understandings. We've been so engaged in rounding things off, smoothing things over, keeping the lid on (the metaphors are numerous), that our organizations have been dying, literally, for information they could feed on, information that was different, disconfirming, and filled with enough instability to knock the system into new life.

We do not exist at the whim of random information; that is not the fearsome prospect which greets us in conscious organizations. Our own consciousness plays a crucial role. We, alone and in groups, serve as gatekeepers, deciding which fluctuations to pay attention to, which to suppress. We already are highly skilled at this, but the gate-keeping criteria need revision. We need to open the gates to more information, in more places, and to seek out information that is ambiguous, complex, of no immediate value. I know of one organization that thinks of information as a salmon. If its organizational streams are well-stocked, the belief goes, information will find its way to where it needs to be. The organization's job is to keep the streams clear, so that the salmon have an easy time of it. The result is a harvest of new ideas and projects.

Information is always spawned out of uncertain, even chaotic circumstances. This is not a reassuring prospect. How are we to welcome information into our organizations and ally ourselves with it as a partner in our search for organizational order, if the processes that give it birth are ambiguity and complexity? In a profession that has raised the practice of "no surprises" to a high art, sponsoring such processes reads like a macabre prescription for self-destruction. Few things make us more frantic than increasing complexity. And although we say we've come to tolerate ambiguity rather well over the past years (because we had no other choice—it wasn't going away), it often appears that we don't tolerate it as much as we shield ourselves from it. We have a hard time with lack of clarity, or with questions that have no readily available answers. We quickly find our way out of these discomforts, focusing on one element, coming up with a solution, and pretending not to notice the questions we've left hanging. We feel safer with blinders on, fearing that unimpaired vision will only add to our distress.

We fear both ambiguity and complexity in management because we still focus on the parts, rather than the whole system. We still believe that influence is a localized event, where we must directly touch what we seek to affect. We still believe that what holds a system together are point to point connections that must be laboriously woven together by us. Complexity only adds to our task, requiring us to keep track of more things, handle more pieces, make more connections. As things increase in number or detail, the span of control stretches out elastically, and, suddenly, we are snapped into unmanagability.

But there is a way out of this fear of complexity, and we find it as we step back and refocus our attention on the whole. When we give up myopic attention to details and stand far enough away to observe the movement of the total system, we develop a new appreciation for what is required to manage a complex system. Peter Senge, in his work in systems theory (1990), develops complex nonlinear systems to portray the dynamics of an organization. This whole-system view requires very different management expectations and analytic processes. Rather than creating a model that forecasts the future of the system, nonlinear models encourage the modeler to play with them and observe what happens. Different variables are tried out "in order to learn about the system's critical points and its homeostasis," Senge reports. *Controlling* the model is neither a goal nor an expectation. Analysts want to increase their intuitions about how the system works so they "can interact with it more harmoniously" (in Briggs and Peat 1989, 175).

This is such a remarkably different approach to analysis, this sensing into the movement and shape of a system, this desire to be in harmony with it. The more we develop a sensitivity to systems, the more we redefine our role in managing the system. The intent is not to find the one variable or set of variables that will allow us to assert control. This has always been an illusion anyway. Rather, the intent becomes one of understanding movement based on a deep respect for the web of activity and relationships that comprise the system. Physicist David Peat terms this "gentle action . . . involving extremely subtle actions that are widely distributed over the whole system." The intent is not to push and pull, but rather to give form to what is unfolding (1991, 217-220).

A system's perspective, then, can handle complexity because it does not need to deal with it in a linear fashion. We don't need to make point-to-point connections among separate things; we don't need to move information along linear pathways. Managers have long treated information this way, guiding it through channels, passing it onto the next point. We've been inspired in this by mechanistic models of brain function, believing information is assiduously moved along neural pathways, passed from one neuron to the next. But we are beginning to understand brain function differently.

Newer theories of the brain describe information as widely distributed, not necessarily limited to specific neuron sites. In mapping areas of the brain to determine those that relate to specific signals (for example, those related to hand movements), neuroscientists have found that these "sites" do not correspond to any particular neurons. Instead of a specific physical place, researchers have observed a more fluid pattern of electrical activity. Instructions, such as those for a particular finger movement, seem to be distributed through a shifting network. These memories, it is now thought, "must arise as relationships within the whole neural network" (Briggs and Peat 1989, 171). Where information is stored in networks of relationships among neurons, damage to a particular area of the brain will not result in the loss of that information. Other areas in the network may retain that information in some form.

These *neural nets* have been simulated to some degree in computers using parallel processing. Zohar describes them as a "rather messy, higgledy-piggledy wiring design, where everything seems randomly connected to everything else" (1990, 72). In both of these systems—our brains and the computers that mimic them—complex information travels seemingly randomly across broad expanses, organizing into memory and functions.

Instead of channeled flows of information, we have images of neural nets transmitting information in all directions simultaneously. How this rather "higgledy-piggledy" system works is not clear. We can neither precisely track nor control how such random distribution of information achieves a sense-making capacity. But we each live with the evidence of its effectiveness.

In a hologram, every part contains enough information, in condensed form, to display the whole. "The part is in the whole and the whole is in the part . . . ; *the part has access to the whole*," writes scientist and science commentator Ken Wilbur (1985, 2; italics added). When light is reflected from an object, it creates wave patterns based on the light scattered by the object. These wave patterns are stored on a photographic plate as interference patterns. The image looks blurred, even random. But when a laser light is shone on the image, the original wave pattern is regenerated and what emerges is a three-dimensional image of the whole object. The image of the whole can be reconstructed from any fragment of the original image.

Holograms create wonderful images for the distribution of information in organizations. In fact, we already have an experience with organizational holograms in our current approach to customer service. Most organizations acknowledge that when a customer comes in contact with *anyone* from the organization, no matter his or her position, the customer experiences the *total organization*, for good or ill. Under the laser light of these "moments of truth" (Jan Carlzon of SAS's phrase), the organization becomes visible. We can improve the image that is regenerated by the glare of customer scrutiny only if we understand that every employee has these holographic qualities and truly is capable of reflecting back the image of the total organization. We improve customer satisfaction when we recognize and support organizations as holograms. Just like an actual hologram, if we distribute information broadly across the organization, we strengthen its image.

We have other models in our experience that teach us about the benefits of creating complex levels of information in organizations. The literature on organizational innovation is rich in lessons that apply here; and, not surprisingly, it describes processes that are also prevalent in the natural universe. Innovation is fostered by information gathered from new connections; from insights gained by journeys into other disciplines or places; from active, collegial networks and fluid, open boundaries. Innovation arises from ongoing circles of exchange, where information is not just accumulated or stored, but created. Knowledge is generated anew from connections that weren't there before. When this information self-organizes, innovations occur, the progeny of information-rich, ambiguous environments.

The process of information generating and then self-organizing is evident in one type of planning model that I and others use, that of a future search conference (see Weisbord 1987, ch. 14; and forthcoming). The purpose is to get the whole system in the room to develop a desired future for the organization. People from all parts of the organization as well as outside constituents work together, generating information on the organization's past experiences, internal capacities, and external demands. The first days are spent bringing to the surface the information contained in the organizational neural net of the people in the room. Information is generated in deliberately overwhelming amounts. But by the end of two or three days, the group self-organizes, weaving all that information into potent visions of the future. Rather than basing agreements on the lowest common denominator, the organization present at the conference has self-organized into a higher form, with new and challenging directions.

Although in futures search work complex levels of information are intentionally created, in nature it doesn't take much information, necessarily, to create interesting new structures. Simple information can complexify into new forms just from being fed back on itself. One result is displayed in the ineffable beauty of fractals (see pages 81-81). These geometrical forms are generated by computers from relatively little information expressed in as few as three nonlinear equations. When the equations are fed back on themselves—a process of "evolving feedback"—elaborate levels of differentiation and scaling are created.

> Fractals are . . . complex by virtue of their infinite detail and unique mathematical properties (no two fractals are the same), yet they're simple because they can be generated through successive applications of simple iterations. . . . It's a new brand of reductionism . . . utterly unlike the old reductionism, which sees complexity as built up out of simple forms, as an intricate building is made out of a few simple shapes or bricks. *Here the simple iteration in effect liberates the complexity hidden within it, giving access to creative potential.* The equation isn't the plot of a shape as it is in Euclid. Rather, the equation provides the starting point for evolving feedback. (Briggs and Peat 1989, 104; italics added)

As a consultant, the most important intervention I ever make is when I feed back organizational data to the whole organization. The data often are quite simple, containing a large percentage of information that is already known to many in the organization.

But when the organization is willing to give public voice to the information—to listen to different interpretations and to process them together—the information becomes amplified. In this process of shared reflection, a small finding can grow as it feeds back on itself, building in significance with each new perception or interpretation. As with the creation of fractals, the simple process of iteration eventually reveals the complexity hidden within the issue. From this level of understanding, creative responses emerge and significant change becomes possible.

Our search for organizations that are well-ordered by open and flowing information leads to two complementary processes and tasks: those that create new information, and those that feed existing information back on itself. We already know many of these processes; we just need to emphasize them differently or give them more freedom in their workings. For example, information can be created every time we bring people together in new ways. Activities that create circulation and movement, even the old chestnuts of work teams, job rotations, and task forces, are all potential creators of information. We often limit their potential because we circumscribe them with rules and chains of command or give them narrow mandates or restrict their access to information. But if we liberate them from those confines and allow them greater autonomy, constrained more by purpose than by rules or preset expectations, then their potential for generating information is great.

We also create order when we invite conflicts and contradictions to rise to the surface, when we search them out, highlight them, even allowing them to grow large and worrisome. We need to support people in the hunt for unsettling or discomfirming information, and provide them with the resources of time, colleagues, and opportunities for processing the information. We've seen the value of this process in quality programs and participative management. In such companies, workers are encouraged to look for fluctuations, and processes are in place to support discussions among many levels of the organization. Through constant exchanges, new information is spawned, and the organization grows in effectiveness. I am intrigued by the thought that these programs work well, not simply because they support employee contribution and involvement, but because they generate the very energy that orders the universe—information.

We can encourage vital organizational ambiguity with plans that are open, visions that inspire but do not describe, and by the encouragement of questions that ask "Why?" many times over. Jantsch asks managers to be "equilibrium busters." No longer the caretakers of order, we become the facilitators of disorder. We stir things up and roil the pot, looking always for those disturbances that challenge and disrupt until, finally, things become so jumbled that we reorganize work at a new level of efficacy.

If we accept this challenge to be equilibrium busters, we will find the task easier than we had thought. Complexity is achieved easily these days simply because so much information is available in non-linear, diverse forms. Our thinking processes have always yielded riches when we've approached things openly, letting free associations form into new ideas. Many would argue that we've used such a small part of our mental capacity because of our insistence on linear thinking. Now we have the technology to mirror more generative processes. More and more, the world of information is associative, networked, and heuristic. We are coming to understand the importance of relationships and non-linear connections as the source of new knowledge. Our task is to create organizational forms that facilitate these processes.

Gore Associates, manufacturers of GoreTex®, models one such structure with its open "lattice organization." Roles and structure are created from need and interest; relationships, exchanges, and connections among employees (almost everyone bears the title *associate*) are nurtured as the primary source of organizational creativity and success. One observer noted that the issue was not who or what position would take care of the problem, but what energy, skill, influence, and wisdom were available to contribute to the solution (Pacanowski 1988).

Slowly but perceptibly, other organizations are moving into the realm of increased consciousness. Thinking has become a precious resource, and not just at higher levels of management. We now recognize that many workers need to be trained to interpret the interactions among complex variables. "Intellectual capital" is on the rise, a phrase that tells of the new value being placed on the capacity to generate knowledge. More and more, there is an openness to inter- and intra-organizational exchanges, to decreasing layers of hierarchy, to smart machines, and to the flow of information among all levels. Learning organizations are taking hold. Consciousness is growing. Is new order on the way?

My own faith in the evolution of organizations to higher levels of consciousness arises from my growing understanding and belief that this is an intrinsically well-ordered universe. As I read further into biology and physics, I recognize that natural systems engage with the universe differently than we do.

We struggle to build layer upon layer, while they unfold. We labor hard to hold things together, while they participate openly and complex structures emerge. Jantsch contrasts these two approaches: "[Building-up] emphasizes structure and describes the emergence of hierarchical levels by the joining of systems 'from the bottom up.' Unfolding, in contrast, implies the interweaving of processes which lead simultaneously to phenomena of structuration at different hierarchical levels. . . . Complexity emerges from the interpenetration of processes of differentiation and integration, processes running 'from the top down' and 'from the bottom up' at the same time" (1980, 75).

We need to learn more about this "interweaving of processes" that leads to structure. In ways we have never noticed, the whole of a system manages itself as a *total* system through natural processes that maintain its integrity. It is critical that we see these processes. It will shift our attention away from the *parts*, those rusting holdovers from an earlier age of organization, and focus us on the deeper, embedded processes that create whole organizations. "What is needed," writes Bohm, "is an *act of understanding* in which we see the totality as an actual process that, when carried out properly, tends to bring about a harmonious and orderly overall action . . . in which analysis into parts has no meaning" (1980, 56).

In quantum physics, *relational holism* describes how whole systems are created among the subatomic particles. In this process, the parts are forever changed, drawn together by a process of internal connectedness. Electrons are drawn into these intimate relations as their wave aspects interfere with one another, overlapping and merging; their own qualities of mass, charge, spin, position, and momentum become indistinguishable from one another. "The whole will, as a whole, possess a definite mass, charge, spin, and so on, but it is completely indeterminate which constituent electrons are contributing what to this. Indeed, it is no longer meaningful to talk of the constituent electrons' individual properties, as these continually change to meet the requirements of the whole" (Zohar 1990, 99).

This is an intriguing image for organizations. It is not difficult to recognize the waves we create in organizations, how we move, merging with others, forming new wholes, being forever changed in the process. We experience this when we say that a team has "jelled," suddenly able to work in harmony, the ragged edges gone, a pleasurable flow to the work. We all have experienced things "coming together," but it has always felt slightly miraculous. We never understood that we were participants in a universe that thrives on information and that will work with us in the creation of order.

Much of the present thinking about organizational design stresses fluid and permeable forms that can be resilient to unending change. These ideas have sparked both hesitation and curiosity. Perhaps if we understand the deep support we have from natural processes, it will help dispel some of the fear. It is not that we are moving toward disorder when we dissolve current structures and speak of worlds without boundaries. Rather, we are engaging in a fundamentally new relationship with order, order that is identified in processes that only temporarily manifest themselves in structures. Order itself is not rigid, but a dynamic energy swirling around us. Relational holism and self-organization work in tandem to give us the living universe. Two dynamic processes, fed by information, combine to create an ordered world. The result is evolution, the organization of information into new forms. Life goes on, richer, more creative than before.

"Thus before all else, there came into
being the Gaping Chasm, Chaos, but there
followed the broad-chested Earth, Gaia, the
forever-secure seat of the immortals . . .
and also Love, Eros, the most beautiful
of the immortal gods, he who breaks limbs."

—Hesiod

CHAPTER 7

Chaos and the Strange Attractor of Meaning

Several thousand years ago, when primal forces haunted human imagination, great gods arose in myths to explain the creation of all beings. At the very beginning was Chaos, the endless, yawning chasm devoid of form or fullness, and Gaia, the mother of the earth who brought forth form and stability. In Greek consciousness, Chaos and Gaia were partners, two primordial powers engaged in a duet of opposition and resonance, creating everything we know.

These two mythic figures again inhabit our imaginations and our science. They have taken on new life as scientists explore more deeply the workings of our universe. I find this return to mythic wisdom both intriguing and comforting. It signifies that a new relationship with Chaos is available, even in the midst of increasing turbulence. Like ancient Gaia, we need to appreciate the necessity for Chaos, understanding it as the life source of our creative power. From his great chasm comes support and opposition, creating the "light without which no form would be visible" (Bonnefoy 1991, 369-70). We, the generative force, give birth to form and meaning, dispelling Chaos with our creative expression. We fill the void with worlds of our creation and turn our backs on him. But we must remember, so the Greeks and our science tell us, that deep within our Gaian centers lives always the dark heart of Chaos.

The heart of chaos has been revealed with modern computers. Watching chaos emerge on a computer screen is a mesmerizing experience. The computer tracks the evolution of a system, recording a moment in the system's state as a point of light on the screen. With the speed typical of computers, we can soon observe millions of moments in the system's history. The system careens back and forth with violent unpredictability, never showing up in the same spot twice. This chaotic movement is seen as rapidly moving lines zooming back and forth across the screen. But as we watch, the lines weave their strands into a pattern, and an order to this disorder emerges. The chaotic movements of the system have a shape. The shape is a "strange attractor," and what has appeared on the screen is the order inherent in chaos (see illustration, page 86).

Chaos has always had a shape—a concept contradictory to our common definition of chaos—but until we could see it through the eyes of computers, we saw only turbulence, energy without predictable form or direction. Chaos is the final state in a system's movement away from order. Not all systems move into chaos, but if a system is dislodged from its stable state, it moves first into a period of oscillation, swinging back and forth between different states. If it moves from this oscillation, the next state is full chaos, a period of total unpredictability. But in the realm of chaos, where everything should fall apart, the strange attractor comes into play. (Science uses other attractors. These particular ones were named *strange* by two scientists, David Ruelle and Floris Takens, who felt the name was deeply suggestive [Gleick 1987, 131]. As Ruelle said, "The name is beautiful and well-suited to these astonishing objects, of which we understand so little [in Coveney and Highfield 1990, 204].)

A strange attractor is a basin of attraction, an area displayed in computer-generated phase space that the system is magnetically drawn into, pulling the system into a visible shape. Computer phase space is multi-dimensional, allowing scientists to see a system's movement in more dimensions than had been possible previously. Shapes that were not visible in two dimensions now become apparent. In a chaotic system, scientists now can observe movements that, though random and unpredictable, never exceed finite boundaries. "Chaos," says planning expert T. J. Cartwright, "is order without predictability" (1991, 44). The system has infinite possibilities, wandering wherever it pleases, sampling new configurations of itself. But its wandering and experimentation respect a boundary.

Ruelle, like many chaos scientists, reaches for poetic language to describe these strange attractors: "These systems of curves, these clouds of points, suggest sometimes fireworks or galaxies, sometimes strange and disquieting vegetal proliferations. A realm lies there of forms to explore, of harmonies to discover" (in Coveney and Highfield 1990, 206).

Briggs and Peat, in describing the computer images of systems wandering between orderly and chaotic states, paint a similarly compelling picture of this dance between turbulence and order:

> Evidently familiar order and chaotic order are laminated like bands of intermittency. Wandering into certain bands, a system is extruded and bent back on itself as it iterates, dragged toward disintegration, transformation, and chaos. Inside other bands, systems cycle dynamically, maintaining their shapes for long periods of time. But eventually all orderly systems will feel the wild, seductive pull of the strange chaotic attractor. (1989, 76-77)

In much of new science, we are challenged by paradoxical concepts—matter

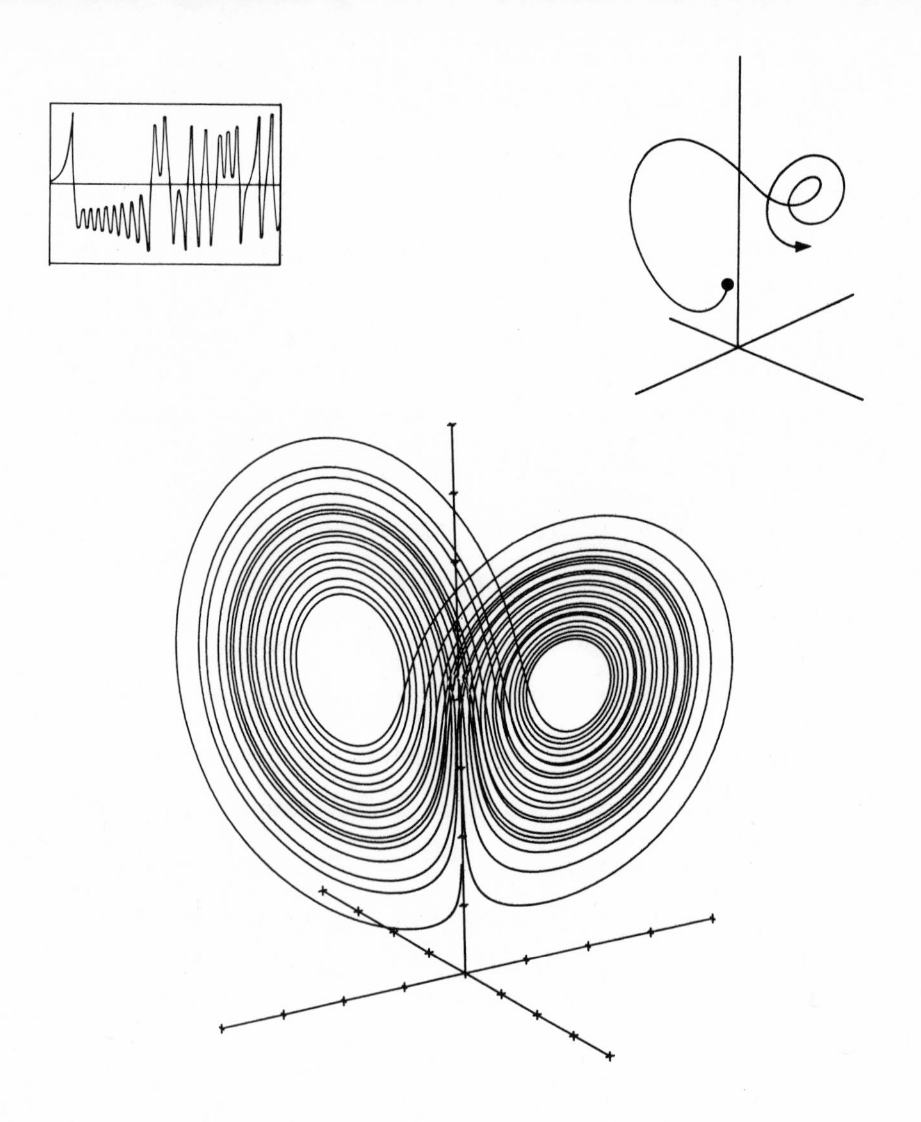

This wonderful, well-ordered butterfly or owl-shaped image of a chaotic system was not visible to scientists until they developed a way to plot the development of a system using multiple variables. Traditional plots of one variable (upper left) show a system in chaos—total unpredictability. However, in phase space, three variables are plotted simultaneously; as the system wanders chaotically, the location of the system can be plotted in three-dimensional space (upper right). This perspective shows the emergence of a strange attractor, the boundaries that contain chaos. The system never lands in the same place twice, yet it never exceeds certain boundaries. As the attractor takes shape, it contains layer upon layer of trajectories that never intersect.

From Gleick 1987, used with permission

that is immaterial, disequilibrium that creates global equilibrium, and now chaos that is non-chaotic. Yet the paradox of chaos was known anciently, in its mythic pairing with order. In every system lurks the potential for chaos, "a creature slumbering deep inside the perfectly ordered system" (Briggs and Peat 1989, 62). But chaos, when it erupts, will never exceed the bounds of its strange attractor. This mirror world of order and disorder challenges us to look, once again, at the whole of the system. Only when we step back to observe the shape of things can we see the patterns of movement from chaos to order and from order to chaos. "Here," recalls chaos scientist Doyne Farmer, "was one coin with two sides. Here was order, with randomness emerging, and then one step further away was randomness with its own underlying order" (in Gleick 1987, 252).

Chaos of this nature (known as deterministic chaos) is created by iterations in a non-linear system, information feeding back on itself and changing in the process. (This process of iteration also characterizes the self-organization observed in biological and chemical systems [see chapter 5].) Non-linearity has been described by Coveney and Highfield as "getting more than you bargained for" (1990, 184). Very slight variances in the conditions of the equation, variances so small as to be indiscernible, amplify into unpredictable results when they are fed back on themselves. If the system is non-linear, iterations can take the system in any direction, away from anything we might expect. The proverbial straw that broke the camel's back is one familiar example of non-linearity: A very small change had an impact far beyond what could have been predicted.

Until recently, we discounted the effects of non-linearity, even though it abounds in life. We had been trained to believe that small differences averaged out, that slight variances converged toward a point, and that approximations would give us a fairly accurate picture of what could happen. But chaos theory ended all that. In a dynamic, changing system, the *slightest* variation can have explosive results. Hypothetically, were we to create a difference in two values as small as rounding them off to the thirty-first decimal place (calculating numbers this large would require a computer of astronomical size), after only one hundred iterations the whole calculation would go askew. The paths of the two systems would have diverged in unpredictable ways. Even infinitesimal differences are far from inconsequential. "Chaos takes them," physicist James Crutchfield says, "and blows them up in your face" (in Briggs and Peat 1989, 73).

Scientists now emphasize the very small differences at the beginning of a system's evolution that make prediction impossible; this is termed "sensitive dependence on initial conditions." Edward Lorenz, a meteorologist, first drew modern-day attention to this with his "butterfly effect." (At the end of the nineteenth century, mathematician, physicist, and philosopher Henri Poincaré had called attention to chaos in dynamic systems and its impact on prediction, but it was chaos science late in this century that revived his findings.) Does the flap of a butterfly wing in Tokyo, Lorenz queried, affect a tornado in Texas (or a thunderstorm in New York)? Though unfortunate for the future of accurate weather prediction, his answer was "yes."

Science has been profoundly affected by this new relationship with the non-linear character of our world. Many of the prevailing assumptions of scientific thought have had to be recanted. As the scientist Arthur Winfree expresses it:

> The basic idea of Western science is that you don't have to take into account the falling of a leaf on some planet in another galaxy when you're trying to

> account for the motion of a billiard ball on a pool table on earth. Very small influences can be neglected. There's a convergence in the way things work, and arbitrarily small influences don't blow up to have arbitrarily large effects. (In Gleick 1987, 15)

But chaos theory has proved these assumptions false. The world is far more sensitive than we had ever thought. We may harbor the hope that we will regain predictability as soon as we can learn how to account for all variables, but in fact no level of detail can ever satisfy this desire. Iteration creates powerful and unpredictable effects in non-linear systems. In complex ways that no model will ever capture, the system feeds back on itself, enfolding all that has happened, magnifying slight variances, encoding it in the system's memory—and prohibiting prediction, ever.

Chaos theory is based on Newtonian mechanical principles, but in its unpredictability, it shares the uncertainty experienced at the quantum level. In both sciences, uncertainty arises because the *wholeness* of the universe resists being studied in pieces. Briggs and Peat, in their intriguing exploration of the mirror world of chaos and order, suggest that wholeness is "what rushes in under the guise of chaos whenever scientists try to separate and measure dynamical systems as if they were composed of parts. . . . The whole shape of things depends upon the minutest part. The part *is* the whole in this respect, for through the action of any part, the whole in the form of chaos or transformative change may manifest" (1987, 74-75). The strange attractors that form on our screens, Briggs and Peat suggest, are not the shape of chaos. They are the shape of wholeness.

Iteration launches a system on a journey that visits both chaos and order. The most beautiful images of iteration are found in the artistry of fractals, computer-generated models drawn by the iteration of a few equations (see also chapter 6). The equations change as they are fed back on themselves. After countless iterations, their tracks materialize into form, creating detailed shapes at finer and finer levels. Everywhere in this minutely detailed fractal landscape, there is self-similarity. The shape we see at one magnification we will see at all others. No matter how deeply we look, peering down through very great magnifications, the same forms are evident. There is pattern within pattern within pattern. There is no end to them, no scale small enough that these intricate shapes cease to form. Because the formations go on forever, there is no way to ever gain a finite measurement of them. We could follow the outline of forms forever, and at ever smaller levels, there would always be something more to measure (see illustrations, pages 80-81).

Fractals entered our world through the research of Benoit Mandelbrot of IBM. In discovering them, he gave us a language, a form of geometry, that allowed us to understand nature in new ways. Fractals are everywhere around us, in the patterns by which nature forms clouds, landscapes, circulatory systems, trees, and plants. We observe fractals daily, but until recently, we lacked a means for understanding them or how they were created.

Fractals, as common as they are, teach some new and important things. For example, it is impossible to ever know the precise measurement of a fractal. Mandelbrot's seminal fractal exercise was a simple question posed to colleagues and students. "How long is the coast of Britain?" As his colleagues soon learned, there is no final answer to this question. The closer you zoom in on the coastline, the more there is to measure.

Since there can be no definitive measurement, what is important in a fractal landscape is to note the *quality* of the system—its complexity and distinguishing shapes, and how it differs from other fractals. If we ignore these qualitative factors and focus on quantitative measures, we will always be frustrated by the incomplete and never-ending information we receive. Fractals, in stressing *qualitative* measurement, remind us of the lessons of wholeness we encountered in the systems realm. What we *can* know, and what is important to know, is the shape of the whole—how it develops and changes, or how it compares to another system.

In organizations, we are very good at measuring activity. In fact, that is primarily what we do. Fractals suggest the futility of searching for ever finer measures of discrete parts of the system. There is never a satisfying end to this reductionist search, never an end point where we finally know everything about even one part of the system. When we study the individual parts or try to understand the system through its *quantities*, we get lost in a world we can never fully measure nor appreciate. Scientists of chaos study shapes in motion; if we were to approach organizations in a similar way, what would constitute the shape and motion of an organization?

We have started edging toward an answer to this question in our growing focus on studying organizations as whole systems rather than our old focus on discrete tasks. Organizations that are using complex system modeling (mentioned also in chapters 2 and 4) are experimenting with these skills. In other organizations, this newer awareness of dynamic shape may occur simply in how problems are approached. Is there an attempt to step back from the problem, to gain enough perspective so that its shape emerges out of the myriad variables that influence it? Are people encouraged to look for themes and patterns, rather than isolated causes? Some of the analytic tools introduced in corporate quality programs, although relying initially on diverse and minute mathematical information, eventually prove effective because they allow people to appreciate the complex and ever-changing shape of the organization, and how multiple forces work together to form it.

Fractal principles have given us valuable insight into how nature creates the shapes we observe. Mountains, rivers, coastlines, vegetables, lungs, circulatory systems—all of these (and millions more) are fractal, replicating a dominant pattern at several smaller levels of scale (see illustrations, pages 80-81.) The scientist Michael Barnsley was intrigued to see if he could recreate the shapes of natural objects by deducing the initial equations that described their forms. He invented the "Chaos Game," in which he decoded objects to derive the number rules that expressed certain global information about their shapes (his first attempt was with a fern). These numbers captured only essential information about the shape. They were surprisingly simple, devoid of the levels of precise prescriptive information we might expect. Barnsley then interjected randomness, setting the numbers loose to feed back on themselves. They were allowed to follow their own iterative wanderings, working at whatever scale they chose. With this approach, he successfully reproduced an entire computer garden of plant shapes (see fern illustration, page 131).

His work with "random fractals" and the chaos game are very instructive. First, Barnsley shows us that predictability still exists. The shapes that he created are predictable, built into the numbers. But indeterminism (randomness) also plays a key role. It is randomness that leads to the creation of the pattern at different levels of scale. The same shapes appear predictably everywhere with only simple levels of instruction and large amounts of freedom. It seems that with a few simple guidelines, left to develop and change randomly, nature creates the complexity and harmony of form we see everywhere.

Many disciplines have seized upon fractals, testing whether self-similar phenomena occur at different levels of scale in both natural and man-made

---

This computer-generated fern is the product of Michael Barnsley's Chaos Game. This remarkable imitation is created by random iterations of a few simple rules that work together to describe the overall shape of the fern. The level of detail that emerges from such simplicity is the result of chaos and order working in concert together.

systems. For example, business forecasters and stock analysts have observed a fractal quality to stock market behaviors and have seen patterns that resemble one another in daily and monthly market fluctuations.

And I believe that fractals also have direct application for the leadership of organizations. The very best organizations have a fractal quality to them. An observer of such an organization can tell what the organization's values and ways of doing business are by watching anyone, whether it be a production floor employee or a senior manager. There is a consistency and predictability to the quality of behavior. No matter where we look in these organizations, self-similarity is found in its people, in spite of the complex range of roles and levels.

How is this quality achieved? The potent force that shapes behavior in these fractal organizations, as in all natural systems, is the combination of simply expressed expectations of acceptable behavior and the freedom available to individuals to assert themselves in non-deterministic ways. Fractal organizations, though they may never have heard the word *fractal*, have learned to trust in natural organizing phenomena. They trust in the power of guiding principles or values, knowing that they are strong enough influencers of behavior to shape every employee into a desired representative of the organization. These organizations expect to see similar behaviors show up at every level in the organization because those behaviors were patterned into the organizing principles at the very start.

Fractals and strange attractors echo the principles evidenced in the globally stable, locally changing structures we observed in self-organizing systems. In both realms, whether it be a biological system or a mathematical rendering of a chaotic system, the structure is capable of maintaining its overall shape and a large degree of independence from the environment because each part of the system is free to express itself within the context of that system. Fluctuations, randomness, and unpredictability at the local level, in the presence of guiding or self-referential principles, cohere over time into definite and predictable form. It was this odd combination of predictability and self-determination that attracted some early scientists of chaos. The science seemed to explain how free will could be expressed and have value in an orderly universe. "The system is deterministic, but you can't say what it's going to do next" (Gleick 1987, 251).

These ideas speak with a simple clarity to issues of effective leadership. They bring us back to the importance of simple governing principles: guiding visions, strong values, organizational beliefs—the few rules individuals can use to shape their own behavior. The leader's task is to communicate them, to keep them ever-present and clear, and then allow individuals in the system their random, sometimes chaotic-looking meanderings.

This is no simple task. Anytime we see systems in apparent chaos, our training urges us to interfere, to stabilize and shore things up. But if we can trust the workings of chaos, we will see that the dominant shape of our organizations can be maintained if we retain clarity about the purpose and direction of the organization. If we succeed in maintaining focus, rather than hands-on control, we also create the flexibility and responsiveness that every organization craves. What leaders are called upon to do in a chaotic world is to shape their organizations through concepts, not through elaborate rules or structures.

Ever since my imagination was captured by the phrase "strange attractor," I have wondered if we could identify such a force in organizations. Is there a magnetic force, a basin for activity, so attractive that it pulls all behavior toward it and creates coherence? My current belief is that we *do* have such attractors at work in organizations and that one of the most potent shapers of behavior in organizations, and in life, is *meaning*. Our main concern, writes Viktor Frankl in his presentation of logotherapy, "is not to gain pleasure or to avoid pain but rather to see a meaning in . . . life" (1959, 115).

I became aware of the call of meaning in our organizational lives when I worked with a number of incoherent companies that had been tipped into chaos by reorganizations or leveraged buyouts. They had lost any purpose beyond the basic struggle to survive. Yet under these circumstances, I saw some employees who continued to work hard and contribute to the organization even when the organization could offer them nothing, not even the promise of a job in the future. Most employees had, more predictably, checked out psychologically, just putting in their time, waiting for the inevitable. But others stayed creative and focused on creating new services, even with the great uncertainty of the future. This puzzled me greatly.

I assumed at first that they were simply denying reality. But when I talked to these employees, it became evident that something else much more important was going on. They were staying creative, making sense out of non-sense, because they had taken the time to create a meaning for their work, one that transcended present organizational circumstances. They wanted to hold onto motivation and direction in the midst of turbulence, and the only way they could do this was by investing the current situation with meaning. Frankl, in *Man's Search for Meaning*, points out very clearly that meaning saved lives in the concentration camps of Germany. The one thing that can never be taken from us, he writes, is our attitude toward a situation. If we search to create meaning, we can survive and even flourish. In chaotic organizations, I observed just such a phenomenon. Employees were wise enough to sense that personal meaning-making was their only route out of chaos. In some ways, the future of the organization became irrelevant. They held onto personal coherence because of the meaning attractor they created. Maybe the organization didn't make sense, but their lives did.

I have also seen companies make deliberate use of meaning to move through times of traumatic change. I've seen leaders make great efforts to speak forthrightly and frequently to employees about current struggles, about the tough times that lie ahead, and about what they dream of for the future. These conversations fill a painful period with new purpose, giving reasons for the current need to sacrifice and hold on. In most cases, given this kind of meaningful information, workers respond with allegiance and energy.

All of us want so much to know the "why" of what is going on. (How often have you heard yourself or others say, "I just wish they would tell me *why* we're doing this"?) We instinctively reach out to leaders who work with us on creating meaning. Those who give voice and form to our search for meaning, and who help us make our work purposeful, are leaders we cherish, and to whom we return gift for gift.

The formative powers of meaning echo back, at least in my own thinking, to a lesson I learned from self-organizing systems, where the principle of self-reference or self-consistency plays such a critical role. A self-organizing system has the freedom to grow and evolve, guided only by one rule: It must remain consistent with itself and its past. The presence of this guiding rule allows for both creativity and boundaries, for evolution and coherence, for determinism and free will.

When I think about meaning as a strange attractor, I see links to these sciences. Meaning or purpose serves as a point of reference. As long as we keep purpose in focus in both our organizational and private lives, we are able to wander through the realms of chaos, make decisions about what actions will be consistent with our purpose, and emerge with a discernible pattern or shape to our lives.

When a meaning attractor is in place in an organization, employees can be trusted to move freely, drawn in many directions by their energy and creativity. There is no need to insist, through regimentation or supervision, that any two individuals act in precisely the same way. We know they will be affected and shaped by the attractor, their behavior never going out of bounds. We trust that they will heed the call of the attractor and stay within its basin. We believe that little else is required except the cohering presence of a purpose, which gives people the capacity for self-reference.

The science of strange attractors can be linked back, in its images and teachings, to other sciences in many ways. Chaos theory, based on Newtonian mechanics and applicable to the world of large objects, conjures up visions of unseen forces that create order and manage coherence. The fields of quantum space speak of energy that takes form when two subatomic fields intersect. The fields of biological morphogenesis describe physical forms shaped by invisible geometries. It is important to keep the distinctions between these sciences clear (at least for now), but it is also important to note a resonant similarity. Each attempts to describe the presence of nonvisible influences that facilitate orderly processes of creation and change.

In chaos theory it is axiomatic that you can never tell where the system is headed until you've observed it over time. This is also true for organizations, and it is what makes trusting something as ethereal as a strange attractor difficult. It takes time to see if a meaning-rich organization really works. A few are already out there, bright beacons to the future. But if they have not been part of our own experience, we are back to acts of faith. As the universe keeps revealing more of these invisible allies, perhaps we will grow in the belief that systems can evolve into an orderly shape when they center around clear points of self-reference.

We can use our own lives as evidence for this because they evolve in just such a fashion. By the end of our lifetime, we are able to discern our individual basins of attraction. What has been the shape of our life? What has made seemingly random events now appear purposeful? What has made "chance" meetings fit smoothly into the movement of our lives? We discover that we have been influenced by a meaning that is wholly and uniquely our own. We experience a deeper knowledge of the purpose that structured all of our activities, many times invisibly and without our awareness. Whether we believe that we create this meaning in a retrospective attempt to make sense of our lives, or that we discover meaning as the preexistent creation of a purposeful universe, it is, at the end, only meaning that we seek. Nothing else is attractive, nothing else has the power to cohere an entire lifetime of activity. We become like ancient Gaia who boldly embraced the void, knowing that from Chaos' dark depths she would always pull forth order.

"Science affects the way we think together."

—Lewis Thomas

CHAPTER 8

The New Scientific Management

In the history of human thought, a new way of understanding or a new frame for seeing the world often appears seemingly spontaneously in widely separated places or from several disciplines at once. Darwin proposed his theory of evolution at the same time that another researcher, working in Malaysia, published very similar ideas. Physicist David Peat has pointed out that the understanding of light as energy evolved in parallel ways in both art and science over the centuries. The sixteenth-century Dutch school of painters drew light for its effects on interior spaces, depicting how it entered rooms through cracks or under doors or was transformed as it passed through colored glass. At the same time, Sir Isaac Newton was studying prisms and the behavior of light as it passed through small apertures. Two hundred years later, J. M. W. Turner painted light as energy, a swirling power that dissolved into many forms; simultaneously, physicist James C. Maxwell was formulating his wave theory in which light results from the swirling motion of electrical and magnetic fields. When Impressionist painters explored light for its effects on dissolving forms, even painting it as discrete dots, physicists were theorizing that light was made up of minuscule energy packets known as quanta (Peat 1987, 31-32; for a detailed exploration of the synchronicity between art and physics, see Schlain 1991).

Very recently, we have witnessed a similar springing forth of parallel concepts in science and business. AT&T began to advertise a world of electronic networks and connectivity at the same time quantum physicists began communicating in earnest to us lay persons about cosmic interconnectivity. Scientists and businesspeople use surprisingly similar language to describe this new world of interconnections. When Levi Strauss CEO Robert Haas told an interviewer that "we are at the center of a seamless web of mutual responsibility and collaboration . . . , a seamless partnership, with interrelationships and mutual commitments," it was easy to hear the voices of physicists in the background (in Howard 1990, 136).

Also in the past few years, a new way of thinking about organizations has been emerging. Whether they be large corporations, microbes, or seemingly inert chemical structures, we are now interested in learning about any organization's "self-renewing" properties. We are searching for the secrets that contribute to vitality and growth both in nature and in our workplaces.

This relationship between business and science goes back many years. Although in many ways Newtonian thinking unwittingly inspired organizational design, science was brought deliberately into management theory and credited with giving it more validity in the era of "scientific management" in the early years of this century. Frederick Taylor, Frank Gilbreth, and hosts of their followers led the efforts to engineer work, creating time-motion studies for efficiencies and breaking work into discrete tasks that could be done by the most untrained workers. Though we may have left behind some of the rigid, fragmented structures created during that time, we have not in any way abandoned science as the source of most of our operating principles. Planning, measurement, motivation theory, organizational design—each of these and more bears the recognizable influence of science.

A few months ago I was in the audience at a social science conference, listening to colleagues report on their research. In each presentation, I was struck by how "scientific" we in the social sciences strive to be. It's as if we're afraid that we might lose our credibility without our links to math and physics. (William Bygrave, trained as a physicist but now a student of organizations, calls this "physics envy" [1989, 16].)

In one conference session, an organizational theorist drew a long formula on the board that captured, he assured us, all of the relevant variables an employee would use to decide on further education. To be fair to this man, I need to say that all of my professional life I have had a deep aversion to formulaic descriptions of human behavior. But I sat there aghast. There was his long string of variables—separate descriptors interacting in precise, mathematical ways—and here was my brain, filled with my recent readings about fuzzy particles that are nothing but temporary connections in the webs of an interrelated universe, moments of meeting that cannot be captured in predictable ways. I was struck suddenly by the joke of it all. We social scientists are trying hard to be conscientious, using the methodologies and thought patterns of seventeenth-century science, while the scientists, travelling away from us at the speed of light, are moving into a universe that suggests entirely new ways of understanding. Just when social scientists seem to have gotten the science down and can construct strings of variables in impressive formulae, the scientists have left, plunging ahead into the vast "porridge of being" that describes a new reality.

We need to link up once again with the vital science of our times, not just because of our historic relationship, but because, by now, scientific concepts and methods are embedded deep within our collective unconscious. We cannot escape their influence nor deny the images they have imprinted on our minds as the dominant thought structure of our society.

Among its new images, science can encourage us to develop a different relationship with discovery. Nobel Prize winner Sir Peter Medawar said that scientists build "explanatory structures, *telling stories* which are scrupulously tested to see if they are stories about real life" (in Judson 1987, 3). I like this idea of story-tellers. It works well to describe all of us. We are great weavers of tales, outdoing one another around the campfire to see which stories best capture our imaginations and the experiences of our lives. If we can look at ourselves truthfully in the light of this fire and stop being so serious about getting things "right"—as if there were still an objective reality out there—we can engage in life with a different quality, a different level of playfulness. Lewis Thomas explains that he could tell something important was going on in an experimental laboratory by the laughter. Surprised by what nature has revealed, we find that things at first always look startlingly funny. "Whenever you can hear laughter," Thomas says, "and somebody saying, 'But that's *preposterous!*'—you can tell that things are going well and that something probably worth looking at has begun to happen in the lab" (in Judson 1987, 71).

Wouldn't we all welcome more laughter in the halls of management? I would be excited to encounter people delighted by surprises instead of the ones I now meet who are scared to death of them. Were we to become truly good scientists of our craft, we would seek out surprises, relishing the unpredictable when it finally decided to reveal itself to us. Surprise *is* the only route to discovery, the only path we can take if we're to search out the important principles that can govern our work. The dance of this universe extends to all the relationships we have. Knowing the steps ahead of time is not important; being willing to engage with the music and move freely onto the dance floor is what's key.

One of the guiding principles of scientific inquiry is that at all levels, nature seems to resemble itself. For me, the parsimony of nature's laws is further argument why we need to take science seriously. If nature uses certain principles to create her infinite diversity, it is highly probable that those principles apply to human organizations. There is no reason to think we'd be the exception. Nature's predisposition toward self-similarity can be extremely useful. It can even help us evaluate current management practices, providing a guide through the fads and ideas that plague us, directing our attention to those things that have merit at a deeper level. I feel better able to distinguish real nourishment from fast-food guru advice because of my awareness of the directions science is taking. Although I have intimated throughout these essays some of these connections between new science and current management thinking, I'd like to underline a few of them.

To start, there are many critiques offered for the current and growing shift toward participative management. Is this a popular idea that, like so many others, we can wait out, knowing it will pass? Is it based on democratic principles and therefore non-transferable to other cultures? Is it merely a more sophisticated way to manipulate workers? Or is something else going on? For me, quantum physics answered those questions. I believe in my bones that the movement towards participation is rooted, perhaps subconsciously for now, in our changing perceptions of the organizing principles of the universe. This may sound grandiose, but the quantum realm speaks emphatically to the role of participation, even to its impact on creating reality. As physicists describe this participatory universe, how can we fail to share in it and embrace it in our management practices? Will participation go away? Not until our science changes.

The participatory nature of reality has focused scientific attention on relationships. Nothing exists at the subatomic level, or can be observed, without engagement with another energy source. This focus on relationships is also a dominant theme in today's management advice. For many years, the prevailing maxim of management stated: "Management is getting work done through others." The important thing was the work; the "others" were nuisances that needed to be managed into conformity and predictability. Managers have recently been urged to notice that they have *people* working for them. They have been advised that work gets done by humans like themselves, each with strong desires for recognition and connectedness. The more they (we) feel part of the organization, the more work gets done.

This, of course, brings with it a host of new, relationship-based problems that are receiving much notice. How do we get people to work well together? How do we honor and benefit from diversity? How do we get teams working together quickly and efficiently? How do we resolve conflicts? These *relationships* are confusing and hard to manage, so much so that after a few years away from their MBA programs, most managers report that they wish they had focused more on people management skills while in school.

Leadership skills have also taken on a relational slant. Leaders are being encouraged to include stakeholders, to evoke followership, to empower others. Earlier, when we focused on tasks, and people were the annoying inconvenience, we thought about "situational" leadership—how the situation could affect our choice of styles. A different understanding of leadership has emerged recently. Leadership is *always* dependent on the context, but the context is established by the *relationships* we value. We cannot hope to influence any situation without respect for the complex network of people who contribute to our organizations. Is this a fad? Or is it the web of the universe becoming felt in our work lives?

Participation and relationships are only part of our present dilemmas. Here we sit in the Information Age, besieged by more information than any mind can handle, trying to make sense of the complexity that continues to grow around us. Is information anything more than a new and perplexing management tool? What if physicist John Archibald Wheeler is right? What if information is the basic ingredient of the universe? This is not a universe of things, but a universe of the "no-thing" of information, where meaning provides the "software" for the creation of forms (Talbot 1986, 157-58). If the universe *is* nothing more than the invisible workings of information, this could explain why quantum physicists observe connections between particles that transcend space and time, or why our acts of observation change what we see. Information doesn't need to obey the laws of matter and energy; it can assume form or communicate instantaneously anywhere in the information picture of the universe.

In organizations, we aren't suffering from information overload just because of technology, and we won't get out from under our information dilemmas just by using more sophisticated information-sorting techniques. Something much bigger is being asked of us. We are moving irrevocably into a new relationship with the creative force of nature. However long we may drag our feet, we will be forced to accept that information—freely generated and freely exchanged—is our only hope for organization. If we fail to recognize its generative properties, we will be unable to manage in this new world.

This new world is also asking us to develop a different understanding of autonomy. To many managers, autonomy is just one small step away from anarchy.

If we are to use it at all, it must be carefully limited. As one manager wryly commented, "I believe in fully autonomous work, as long as it stops at the level below me." Yet everywhere in nature, order is maintained in the midst of change because autonomy exists at local levels. Sub-units absorb change, responding, adapting. What emerges from this constant flux is that wonderful state of *global* stability. Rather than developing pockets of stability and incrementally building them into a stable organization, nature creates ebbs and flows of movement at all levels. These movements merge into a whole that can resist most of the demands for change at the global level because the system has built into it is so much internal motion.

The motion of these systems is kept in harmony by a force we are just beginning to appreciate: the capacity for self-reference. Instead of whirling off in different directions, each part of the system must remain consistent with itself and with all other parts of the system as it changes. There is, even among simple cells, an unerring recognition of the intent of the system, a deep relationship between individual activity and the whole. Could it be possible that nature is guided by something as familiar as Shakespeare's "To thine own self be true"?

More than any other science principle I've encountered, self-reference strikes me as the most important. It conjures up such a different view of management and promises solutions to so many of the dilemmas that plague us: control, motivation, ethics, values, change. And as an operating principle, it decisively separates living organisms from machines. *Star Trek* popularized an effective method for destroying computers; you program them with a self-referential statement, e. g., "Prove that your prime directive is not your prime directive" (Briggs and Peat 1989, 67). As the logic turns back on itself in unending iterations, a machine will blow its circuits. Zen masters employ the same technique with koans, but they know that human brains are not machines and that we can be challenged to new levels of thinking by self-referential exercises.

Perhaps self-reference is the best tool of all for leaving behind the clocklike world of Newton. We can use self-reference to sort out the living from the dead—giving us the means to identify the open systems that thrive on autonomous iterations from the truly mechanistic things in organizations that do best at equilibrium. But before we can use self-reference, we need to solve a deeper problem. We need to be able to trust that something as simple as a clear core of values and vision, kept in motion through continuing dialogue, can lead to order.

At the risk of sounding antiquatedly reductionist, I want to make one more speculation. If management practice is ever to be simplified into one unifying principle, I believe it will be found in self-reference. It is not only the science I have read that gives me such assurance. When I look at the shape and meaning of my own life, and how it has evolved with change, I understand the workings of this principle in intimate detail. For me, there is no choice but to take the paths new science has marked. Like all journeys, this one moves through both the dark and the light, the terrors of the unknown and the joys of deep recognition. Some shapes and landmarks are already clear. Others wait to be discovered. No one, especially the scientists, can say where the journey is leading. But the association promises to be fruitful, and I can feel the explorer's blood rising in me. I am glad to feel in awe again.

"Wisdom is about living harmoniously
in the universe, which is itself a place of order
and justice that triumphs over chaos and employs
chance for its ultimate purpose."

—Matthew Fox

EPILOGUE

Being Comfortable with Uncertainty

Across the valley, the last colors of this day warm the horizon. Two dimensions move across the land, removing all contours, smoothing purple mountains flat against a rose-radiant sky. A volcano erupted in the Philippines six months ago. Now at every twilight, visiting dust shimmers red in the atmosphere, intensifying the colors of an always intense sky. I sit here, bathed in strange light, embraced by dark magenta mountains.

I move differently in the world these days since travelling in the realms of order and chaos and quantum events. It has become a strange and puzzling place where I cannot rely on what I knew and don't yet feel secured by new sources of confidence. It makes things much more interesting, expecting there to be new ways of working without being able to discern them clearly. In the process of writing this book, of playing with its ideas, and of trying things out with others, I've become aware of how difficult it is *not* to be certain. I've encountered, in myself and others, a desire for these new understandings to translate quickly into reliable and trusty tools and techniques. We are not comfortable with chaos, even in our thoughts, and we want to move out of confusion as quickly as possible.

But the science is helping me understand, among many things, the uses of chaos and its role in self-organization. I think I not only expect chaos now, but I've grown more trusting of it as a necessary stage to greater organization. Recently, I advised a group of students who were taking on an ambitious study of a new subject area, and I noticed a different direction to my advice. They were eager to create a model or framework into which they could slot information. I was intent on letting information do its thing. They wanted to get organized at the start; I wanted them to move into confusion. I urged them to create more information than they could possibly handle. I guaranteed them that at some point the information would self-organize in them, crystallizing into interesting forms and ideas.

Neither they nor I have yet seen the results of their work, but I realize that I have not only the science to support this advice. I also have my own experience. I have worked enough with information—as have you—to recognize how it works with us to organize into new forms. But I needed the science to help me see it. In a similar fashion, I believe each of us can validate many of the strange scientific ideas in this book if we look more carefully at our experiences. What we've been given are some new glasses to help us notice the way things have been working all along.

Ralph Waldo Emerson wrote about life as an ongoing encounter with the unknown and created this image: "We wake and find ourselves on a stair; there are stairs below us which we seem to have ascended, there are stairs above us . . . which go out of sight" (in Eiseley 1978, 214). These stairs of understanding we now are climbing feel different. They are less secure, harder to see, and much more challenging. They require very different things from us.

In our past explorations, the tradition was to discover something and then formulate it into answers and solutions that could be widely transferred. But now we are on a journey of mutual and simultaneous exploration. In my view, all we can expect from one another is new and interesting information. We can *not* expect answers. Solutions, as quantum reality teaches, are a temporary event, specific to a context, developed through the relationship of persons and circumstances. There will be no more patrons, waiting expectantly for our return, just more and more explorers venturing out on their own.

This sounds unnerving—I haven't stopped wanting someone, somewhere to return with the right answers. But I know that my hopes are old, based on a different universe. In this new world, you and I make it up as we go along, not because we lack expertise or planning skills, but because that is the nature of reality. Reality changes shape and meaning because of our activity. And it is constantly new. We are required to be there, as active participants. It can't happen without us and nobody can do it for us.

This is a strange world, and it promises to get stranger. Niels Bohr, who engaged with Heisenberg in those long, nighttime conversations that ended in despair, once said that great innovations, when they appear, seem muddled and strange. They are only half-understood by their discoverer and remain a mystery to everyone else. But if an idea does not appear bizarre, he counseled, there is no hope for it (in Wilbur 1985, 20). So we must live with the strange and the bizarre, even as we climb stairs that we want to bring us to a clearer vantage point. Every step requires that we stay comfortable with uncertainty, and confident of confusion's role. After all is said and done, we will have to muddle our way through. But in the midst of muddle—and I hope I remember this—we can walk with a sure step. For these stairs we climb only take us deeper and deeper into a universe of inherent order.

Suggested Readings

If you are interested in further exploration of any of the topics presented here, I would recommend selecting the books below, which are listed by topic. Books in new science keep appearing regularly; this listing describes my own preferences.

More comprehensive, including several scientific perspectives, although structured around a particular topic

Briggs, John, and F. David Peat. *Turbulent Mirror: An Illustrated Guide to Chaos Theory and the Science of Wholeness*. New York: Harper and Row, 1989.

Capra, Fritjof. *The Turning Point: Science, Society and the Rising Culture*. New York: Bantam Books, 1983.

Cole, K.C. *Sympathetic Vibrations: Reflections on Physics as a Way of Life*. New York: Bantam Books, 1985.

Coveney, Peter, and Roger Highfield. *Arrow of Time: A Voyage Through Science to Solve Time's Greatest Mystery.* New York: Fawcett Columbine, 1990.

Ferris, Timothy. *Coming of Age in the Milky Way.* New York: Doubleday, 1988.

Peat, F. David. *The Philosopher's Stone: Chaos, Synchronicity and the Hidden Order of the World*. New York: Bantam Books, 1991.

Talbot, Michael. *Beyond the Quantum.* New York: Bantam Books, 1986.

Wilber, Ken, ed. *The Holographic Paradigm and Other Paradoxes: Exploring the Leading Edge of Science.* Boston: Shambala, 1985.

Quantum Physics

Capra, Fitjof. *The Tao of Physics*. New York: Bantam Books, 1976.

——*The Turning Point: Science, Society and the Rising Culture*. New York: Bantam Books, 1983,

Gribbin, John. *In Search of Schroedinger's Cat: Quantum Physics and Reality.* New York: Bantam Books, 1984.

Herbert, Nick. *Quantum Reality: Beyond the New Physics*. New York: Anchor Books: New York, 1987.

Talbot, Michael. *Beyond the Quantum.* New York: Bantam Books, 1986.

Toben, Bob, and Fred Allen Wolf. *Space, Time and Beyond*. New York: Bantam Books, 1983.

Zukav, Gary. *The Dancing Wu Li Masters*. New York: Bantam Books, 1979.

Self-Organizing Systems

Jantsch, Erich. *The Self-Organizing Universe*. Oxford: Pergamon Press, 1980.

Nonaka, Ikujiro. "Creating Organizational Order Out of Chaos: Self-Renewal in Japanese Firms." *California Management Review*, Spring 1988, 57-73.

Prigogine, Ilya, and Isabelle Stengers. *Order Out of Chaos*. New York: Bantam Books, 1984.

Chaos Theory

Briggs, John, and F. David Peat. *Turbulent Mirror: An Illustrated Guide to Chaos Theory and the Science of Wholeness*. New York: Harper and Row, 1989.

Gleick, James. *Chaos: Making a New Science*. New York: Viking, 1987.

Discussion of the Philosophical and Methodological Issues Raised by New Science

Bohm, David. *Wholeness and the Implicate Order*. London: Ark Paperbacks, 1980.

Chopra, Deepak. *Quantum Healing: Exploring the Frontiers of Mind and Body Medicine*. New York: Bantam Books, 1989.

Jantsch, Erich. *The Self-Organizing Universe*. Oxford: Pergamon Press, 1980.

Lincoln, Yvonna S., ed. *Organizational Theory and Inquiry: The Paradigm Revolution*. Beverly Hills: Sage, 1985.

Peat, David F. *Synchronicity*. New York: Bantam Books, 1987.

Wilbur, Ken. *The Holographic Paradigm and Other Paradoxes: Exploring the Leading Edge of Science*. Boston: Shambala, 1985.

——ed. *Quantum Questions: Mystical Writings of the World's Great Physicists*. Boston: Shambala, 1985.

Zohar, Danah. *The Quantum Self: Human Nature and Consciousness Defined by the New Physics*. New York: William Morrow and Co., 1990.

Organizational Theory and Practice. (Listed here are just a few of the articles and books that describe organizational structures and practices that move in the directions indicated by new science.)

Barry, David. "Managing the Bossless Team: Lessons in Distributed Leadership." *Organizational Dynamics*, Summer 1991, 31-47.

"Being the Boss." *Inc.*, October 1989, 49-65.

Brassard, Michael. *Memory Jogger Plus: The Seven New Tools of Management*. Methuen, Mass.: Goal/QPC, 1987.

Cartwright, T.J. "Planning and Chaos Theory." *APA Journal* (Winter 1991): 44-56.

Dumaine, Brian. "The Bureaucracy Busters." *Fortune,* June 17, 1991, 37-50.

Bygrave, William. "The Entrepreneurship Paradigm (1): A Philosophical Look at Its Research Methodologies." In *Entrepreneurship Now and Then*. Baylor University, Fall, 1989.

Ferchat, Robert A. "The Chaos Factor." *The Corporate Board*, May/June 1990, 8-12.

Howard, Robert. "Values Make the Company: An Interview with Robert Haas." *Harvard Business Review*, Sept-Oct. 1990, 133-144.

Lincoln, Yvonna S., ed. *Organizational Theory and Inquiry: The Paradigm Revolution*. Beverly Hills, Calif.: Sage, 1985.

Morgan, Gareth. *Images of Organization*. Newbury Park, Calif.: Sage Publications, 1986, particularly chapters 4 and 8.

Nonaka, Ikujiro. "Creating Organizational Order Out of Chaos: Self-Renewal in Japanese Firms." *California Management Review*, Spring 1988.

Prahalad, C.K., and Gary Hamel. "The Core Competence of the Corporation." *Harvard Business Review*, May-June 1990, 79-91.

Radzicki, Michael J. "Institutional Dynamics, Deterministic Chaos, and Self-Organizing Systems." *Journal of Economic Issues,* 24 (March 1990): 57-102.

Semler, Ricardo. "Managing without Managers." *Harvard Business Review*, Sept-Oct., 1989, 76-84.

Senge, Peter. *The Fifth Discipline: The Art and Practice of the Learning Organization*. New York: Doubleday/Currency, 1990.

Vaill, Peter M. *Managing as a Performing Art*. San Francisco: Jossey-Bass, 1989.

Weick, Karl. "Substitute for Corporate Strategy." *The Competitive Challenge: Strategies for Industrial Innovation and Renewal*, edited by D.J. Teece. Cambridge, Mass.: Ballinger Publishing Co., 1987.

Weisbord, Marvin R. *Discovering Common Ground: Strategic Futures Conferences for Improving Whole Systems*. San Francisco: Berrett-Koehler, forthcoming.

——*Productive Workplaces: Managing for Dignity, Commitment and Meaning*. San Francisco: Jossey-Bass, 1987.

BIBLIOGRAPHY

Augros, Robert M., and George N. Stanciu. *The New Story of Science*. New York: Bantam Books, 1984.

Bateson, Gregory. *Mind and Nature*. New York: Bantam Books, 1980.

"Being the Boss." *Inc.*, October 1989, 49-65.

Bellah, Robert N., Richard Madsen, et. al. *Habits of the Heart*. New York: Harper and Row, 1985.

Bohm, David. *Wholeness and the Implicate Order*. London: Ark Paperbacks, 1980.

Bonnefoy, Yves. *Mythologies.* Chicago: University of Chicago Press, 1991.

Brassard, Michael. *The Memory Jogger Plus+: The Seven New Tools of Management.* Methuen, Mass.: GOAL/QPC, 1989.

Briggs, John, and F. David Peat. *Turbulent Mirror: An Illustrated Guide to Chaos Theory and the Science of Wholeness.* New York: Harper and Row, 1989.

Burke, James. Presentation at the 4th Annual Conference on CD-Rom, Anaheim, Calif., 1989.

Bygrave, William. "The Entrepreneurship Paradigm (I): A Philosophical Look at Its Research Methodologies" in *Entrepreneurship Now and Then*. Baylor University, Fall 1989.

Capra, Fritjof. *The Tao of Physics*. New York: Bantam Books, 1976.

——*The Turning Point: Science, Society, and the Rising Culture*. New York: Bantam Books, 1983.

Cartwright, T.J. "Planning and Chaos Theory." *APA Journal* (Winter 1991): 44-56.

Chopra, Deepak. *Quantum Healing: Exploring the Frontiers of Mind and Body Science*. New York: Bantam Books, 1989.

——*The New Physics of Healing*. Boulder, Co.: Sounds True Recording, 1990. Audio cassette.

Cohen, M. D. , J.G. March, and J.P. Olsen. "A garbage can model of organizational choice." *Administrative Science Quarterly*, 17, 1974, 1- 25.

Cole, K. C. *Sympathetic Vibrations: Reflections on Physics as a Way of Life*. New York: Bantam Books, 1985.

Coveney, Peter, and Roger Highfield. *The Arrow of Time: A Voyage Through Science to Solve Time's Greatest Mystery*. New York: Fawcett Columbine, 1990.

Davies P. C. W., and J. Brown. *Superstrings: A Theory of Everything?* Cambridge, U.K.: Cambridge University Press, 1988.

Eiseley, Loren. *The Star Thrower*. San Diego: Harvest/HBJ, 1978.

Ferris, Timothy. *Coming of Age in the Milky Way*. New York: Doubleday, 1988.

Feininger, Andreas. *In a Grain of Sand: Exploring Design by Nature*. San Francisco: Sierra Club Books, 1986.

Fox, Matthew. *Creation Spirituality*. San Francisco: Harper, 1991.

Frankl, Viktor. *Man's Search for Meaning*. Boston: Beacon Press, 1959.

Glass, Leon, and Michael C. Mackey. *From Clocks to Chaos*. New Jersey: Princeton University Press, 1988.

Gleick, James. *Chaos: Making a New Science*. New York: Viking, 1987.

Gribbin, John. *In Search of Schroedinger's Cat: Quantum Physics and Reality*. New York: Bantam Books, 1984.

Handy, Charles. *The Age of Unreason*. Cambridge: Harvard Business School Press, 1989.

Heisenberg, Werner. *Physics and Philosophy*. New York: Harper Torchbooks, 1958.

Herbert, Nick. *Quantum Reality: Beyond the New Physics*. New York: Anchor Doubleday, 1985.

Howard, Robert. "Values Make the Company: An Interview with Robert Haas." *Harvard Business Review*, Sept-Oct. 1990, 133-144.

Jantsch, Erich. *The Self-Organizing Universe*. Oxford: Pergamon Press, 1980.

Kanter, Rosabeth. *The Changemasters*. New York: Simon and Schuster, 1983.

Kanter, Rosabeth Moss. *Men and Women of the Corporation*. New York: Basic Books, 1977.

Lincoln, Yvonna S. , ed. *Organizational Theory and Inquiry: The Paradigm Revolution*. Beverly Hills, Calif.: Sage, 1985.

Lovelock, J. E. *Gaia*. New York: Oxford Univ. Press, 1987.

March, Robert H. *Physics for Poets*. Chicago: Contemporary Books, 1978.

Margalef, Ramon. *Co-Evolution Quarterly* (Summer 1975): 49-66.

Margulis, Lynn, and Dorion Sagan. *Micro-cosmos*. New York: Summit Books, 1986.

Maturana, H., and F. Varela. *Autopoiesis and Cognition: The Realization of the Living*. London: Reidl, 1980.

Meadows, Donella. "Whole Earth Models and Systems." *Co-Evolution Quarterly* (Summer 1982): 98-108.

Nohria, N. "Creating New Business Ventures: Network Organization in Market and Corporate Contexts." Ph.D. diss., MIT, 1988.

Nonaka, Ikujiro. "Creating Organizational Order Out of Chaos: Self-Renewal in Japanese Firms." *California Management Review*, Spring 1988, 57-73.

Pacanowski, Michael. "Communication in the Empowering Organization." In *International Communications Association Yearbook II*. J. A. Anderson, ed. Beverly Hills, Calif.: Sage Publications, 1988, 356-379.

Pagels, Heinz. *The Dreams of Reason*. New York: Bantam Books, 1989.

Peat, F. David. *Synchronicity: The Bridge Between Matter and Mind.* New York: Bantam Books, 1987.

——*The Philosopher's Stone: Chaos, Synchronicity and the Hidden Order of the World.* New York: Bantam Books, 1991.

Peitgen, Heinz-Otto, and Dietmar Saupe, eds. *The Science of Fractal Images*. New York: Springer-Verlag, 1988.

Peters, Tom. *Thriving on Chaos*. New York: Knopf, 1987.

Pinchot, G. *Intrapreneuring*. New York: Harper & Row, 1985.

Prahalad, C. K., and Gary Hamel. "The Core Competence of the Corporation." *Harvard Business Review*, May-June 1990, 79-91.

Prigogine, Ilya. *Omni,* May 1983, 85-121.

——and Isabelle Stengers. *Order Out of Chaos*. New York: Bantam Books, 1984.

"Research Roundup." *Science Digest*, June 1984.

Schlain, Leonard. *Art and Physics: Parallel Visions in Space, Time and Light.* New York: William Morrow and Co., 1991.

Semler, Ricardo. "Managing without Managers." *Harvard Business Review*, Sept-Oct. 1989, 76-84.

Sheldrake, Rupert. *A New Science of Life*. Los Angeles: Jeremy Tarcher, 1981.

——*The Presence of the Past*. New York: Vintage Books, 1988.

——and David Bohm. "Morphogenetic Fields and the Implicate Order," *ReVision* 5 (Fall 1982).

Starbuck, W. H. "Organizations and Their Environments." In M. D. Dunnette, ed. *Handbook of Industrial and Organizational Psychology*. New York: Rand, 1976, 1069-1123.

Talbot, Michael. *Beyond the Quantum*. New York: Bantam Books, 1986.

Thompson, William Irwin. *Imaginary Landscape*. New York: St. Martin's Press, 1989.

Toben, Bob, and Fred Allen Wolf. *Space-Time and Beyond*. New York: Bantam Books, 1983.

Tushman, M., and D. Nadler "Organizing for Innovation." *California Management Review*, Spring 1986, 74-92.

Vaill, Peter. *Managing as a Performing Art*. San Francisco: Jossey-Bass Publishers, 1989.

Weick, Karl. *The Social Psychology of Organization*. New York: Random House, 1979.

——"Substitute for Corporate Strategy." In *The Theoretical Context of Strategic Management*, D. J. Teece, ed. Cambridge, Mass.: Ballinger Publishing Co. , 1987.

Weisbord, Marvin. *Discovering Common Ground: Strategic Futures Conferences for Improving Whole Systems*. San Francisco: Berrett-Koehler, forthcoming.

——*Productive Workplaces*. San Francisco: Jossey-Bass, 1987.

Wilbur, Ken, ed. *The Holographic Paradigm and Other Paradoxes*. Boulder, Colorado: Shambala Press, 1985.

——*Quantum Questions*. Boston: Shambala, 1984.

Wilczek, Frank, and Betsy Devine. *Longing for the Harmonies*. New York: W. W. Norton and Co., 1988.

Wolf, Fred Alan. *Parallel Universes*. New York: Touchstone Books, 1988.

——*Star Wave*. New York: Collier Books, 1984.

——*Taking the Quantum Leap*. New York: Harper and Row, 1989.

Zohar, Danah. *The Quantum Self: Human Nature and Consciousness Defined by the New Physics*. New York: William Morrow and Co., 1990.

Zuboff, Shoshonna. *In the Age of the Smart Machine*. New York: Basic Books, 1988.

Zukav, Gary. *The Dancing Wu Li Masters*. New York: Bantam Books, 1979.

INDEX